改訂
草地学用語集

日本草地学会編

2000

東京
株式会社
養賢堂発行

日本草地学会草地学用語委員会

委員長 清水矩宏　　草地試験場
委　員（五十音順）
　　　　永西　修　　畜産試験場
　　　　大槻和夫　　草地試験場
　　　　神田健一　　草地試験場
　　　　河野憲治　　広島大学
　　　　小松敏憲　　草地試験場
　　　　近藤恒夫　　東北農業試験場
　　　　澤田　均　　静岡大学
　　　　月星隆雄　　農業環境技術研究所
　　　　平田昌彦　　宮崎大学
　　　　福山正隆　　九州大学
　　　　山名伸樹　　生物系特定産業技術研究推進機構
　　　　吉村義則　　草地試験場

改訂にあたって

　旧版の草地学用語集が世に出てから，15年が経過した．旧版の刊行は我が国で開催された第15回国際草地学会議に合わせたものであり，まさに時宜を得たものであった．この用語集は新しい草地学の確立，草地農業の発展のために，重要な基礎づくりを果たしてきたことはいうまでもない．しかし，この間の学術研究の進展はめざましく，新しい技術開発にともなう用語の改廃や新分野の用語の必要性など，時代に即応した用語集の改訂が強く望まれるようになった．このような情勢を受け，平成9・10年の用語委員会において，改訂版の作成が開始された．

　改訂にあたっては，旧版の用語を基礎にしつつ，今までの草地学会誌及び講演要旨の表題からキーワードを抽出し，用語集と重複していない用語を付加した．また，外国文献の解読のために，Forage Abstractのキーワードを抽出して付加した．さらに，各委員の専門分野ごとに必要用語を提案して頂いた．これらの用語を検討対象用語（約1万語に達した）として専門分野別に仕分けをし，各用語について草地学において必要という基準で，専門分野ごとに複数委員の審査を経て，用語委員全員に諮って約5千語を用語として収録した．その結果，旧版とはかなり入れ替わったものとなっている．

　さらに，植物名，病害名，昆虫および動物名についても，昨今の生物多様性に関する研究が草地学分野でも進展することを考えて，拡張する方向で全面的に見直した．植物名には，牧草・飼料作物のほかに雑草，野草も草地および飼料畑に関係するものを網羅している．昆虫および動物についても，従前のリストを整理し，害虫にとどまらず牧草地・飼料畑に発生するものを広く採用した．病害名については，既に草地試験場のホームページに「飼料作物病害図鑑」が公開されているが，それとの整合性を図りつつ主要病害を抽出した．病害名には旧版と同様対象作物も記載している．

　本改訂版の発刊には，ほぼ3年の作業が必要であった．この間，用語委員会は交代時期があったにもかかわらず，アクティブに活動頂けた表記委員は継続して協力をお願いした．本務の合間をぬいながら面倒な作業をやり抜いて頂き感謝の念で一杯である．また，多忙な委員諸氏に代わって，用語のデータベース化，基本的な調査，校正作業等を一手に引き受けて委員長のサポートをして頂いた岡田（現姓小林）恵嬢にお礼を申し上げる．おわりに本用語集の出版を快諾，協力頂いた養賢堂の及川　清社長及び矢野勝也編集部長に深く感謝する．

　　2000年6月

　　　　　　　　　　　　　　日本草地学会
　　　　　　　　　　　　　　　会　長　塩　見　正　衞
　　　　　　　　　　　　　　日本草地学会　草地学用語委員会
　　　　　　　　　　　　　　　委員長　清　水　矩　宏

I 和英の部
Japanese － English

［凡例］
（ ）：省略可
［ ］：代替可
【 】：非常用漢字
《 》：説明あるいは注記
pl. ：複数形

I 和英の部

あ

挨拶行動	greeting behavior
アイソザイム，イソ［同位］酵素	isozyme, isoenzyme
アイソトープ，同位体［元素］	isotope
アイドリング，遊歩《家畜の》	idling
亜鉛	zinc, Zn
青刈り，生草［青刈］給与	soiling, fresh forage feeding, zero grazing
青刈［生草］給与，青刈り	soiling, fresh forage feeding, zero grazing
青刈［生草］収量	fresh [green] yield
青刈飼料	soilage, green fodder
青刈飼料作物	soiling forage crop
亜科	subfamily
赤身肉	lean meat
亜寒帯	subarctic zone
秋肥	autumn manuring, fall dressing
秋作	fall [autumn] crop
秋播き	fall [autumn] seeding [sowing]
秋播性	winter habit
亜極相，サブクライマックス	subclimax
アグロバクテリウム	agrobacterium
アグロフォレストリ	agroforestry
亜高山帯	subalpine zone
あご運動	jaw movement
アシドーシス	acidosis
亜種	subspecies
亜硝酸	nitrous acid
亜硝酸塩，亜硝酸の	nitrite
亜硝酸菌	nitrite forming bacteria
亜硝酸態窒素	nitrite nitrogen
亜硝酸の，亜硝酸塩	nitrite
アスコルビン酸，ビタミンC	ascorbic acid, vitamin C
アズマネザサ型［ネザサ型］草地	Pleioblastus-type grassland
アセチレン還元法	acetylene reduction assay
暖かさの指数	warmth index
圧搾，粉砕	crush
圧搾機	crusher
圧ぺん機	crusher
厚播き	dense [heavier, thick] seeding [sowing]
圧密，締固め	consolidation, compaction
後作	succeeding crop
アニオン	anion
亜熱帯	subtropical zone, subtropics
アブシジン酸	abscisic acid
アポミクシス，無融合生殖	apomixis
アマイド，アミド	amide
アミド，アマイド	amide
アミノ酸	amino acid

[2] あ・い

アミラーゼ	amylase
アメダス	Automated Meteorological Data Acquisition System, AMeDAS
アメニティ，快適性	amenity
荒粉，ひき割り	meal
アリューロン層，こ【糊】粉層	aleurone layer
アルカリ	alkali
アルカリ処理わら【藁】	alkali-treated straw
アルカリ土壌	alkaline soil
アルカロイド	alkaloid
アルコール	alcohol
アルビノ，白子，白色種	albino
アルファルファミール	alfalfa meal
アルミニウム	aluminum, Al
アレル，アレレ，対立遺伝子	allele
アレロパシ，他感作用	alleropathy
暗きょ【渠】排水	underdrainage
暗呼吸	dark respiration
安山岩	andesite
安定同位体［元素］	stable isotope
安定密度	steady density
アントシアン	anthocyan
暗発芽	dark germination
暗反応	dark reaction
アンモニア	ammonia
アンモニア化成作用	ammonification
アンモニア植物	ammonia plant
アンモニア処理もみ【籾】殻	ammoniated rice hull [husk]
アンモニア態	ammonium nitrogen
窒素	
アンローダ	unloader

い

胃	stomach
Eh，酸化還元電位	oxdation-reduction potential, redox potential
イースト，酵母	yeast
胃液	gastric juice
イオウ	sulfur, S
イオン交換樹脂	ion-exchange resin
イオン電極法	ion-electrode method
異核共存体［接合体］	heterokaryon
異核接合体［共存体］	heterokaryon
維管束	vascular bundle
維管束しょう【鞘】	vascular bundle sheath
維管束病	vascular disease
育種，繁殖《動物》	breeding
育種家種子	breeder('s) seed
育成牛	raising cattle
育成牧場	raising farm
育苗	rearing [raising] of seedling
異形，変動	variation
異型［異種，ヘテロ］接合体	heterozygote
移行，転流，転座	translocation
維持行動《採食，休憩などの》	maintenance behavior
維持飼料	maintenance ration
異質性，異種性，不均一［質］性	heterogeneity

異質倍数体	allopolyploid, alloploid
い【萎】縮した, わい【矮】性の	dwarf
異種性, 異質性, 不均一［質］性	heterogeneity
異種［異型, ヘテロ］接合体	heterozygote
異常気象	abnormal weather
異常緑変	virescence
移植	transplanting
移植栽培	transplanting culture
移植実験［試験］	transplant experiment
異数性	heteroploidy, aneuploidy
異数体	heteroploid, aneuploid
イソ［同位］酵素, アイソザイム	isoenzyme, isozyme
遺体, 落葉・落枝	litter
一次休眠	primary dormancy
一次根	primary root
一次生産	primary production
一次遷移	primary succession
一時的草地	temporary grassland
一時的放牧地	temporary pasture
一次分げつ	primary tiller
一代雑種, F_1雑種	F_1 hybrid
一年生（の）, 当年生	annual
一倍体, 半数体, 単相体	haploid
1番刈り	first cutting
1番刈乾草	first cutting hay
1番草	first crop
一毛作	one crop system, single cropping
一様分布	uniform distribution
一穂一列法	ear-to-row method
一般組合せ能力	general combining ability
1本植, 個体植	single [spaced] planting
遺伝	heredity, inheritance
遺伝改良量［獲得量］	genetic gain
遺伝学	genetics
遺伝獲得量［改良量］	genetic gain
遺伝形質	inherited character
遺伝子	gene
遺伝子解析［分析］	gene analysis
遺伝子型	genotype
遺伝子組換え作物	transgenic crop
遺伝子源	gene source
遺伝資源	genetic resources
遺伝子座, 座位	locus
遺伝子（突然）変異	gene mutation
遺伝子発現	gene expression
遺伝子頻度	gene frequency
遺伝子分析［解析］	gene analysis
遺伝子流動	gene flow
遺伝相関	genetic correlation
遺伝的異質性	genetic heterogeneity
遺伝（的）構造	genetic structure
遺伝的多様性	genetic diversity
遺伝的等質性	genetic homogeneity
遺伝的ドリフト［浮動］	genetic drift
遺伝的浮動［ドリフト］	genetic drift
遺伝的変異	genetic variation
遺伝的変異性	genetic variability
遺伝的雄性不念	genetic male sterility
遺伝様式	mode of inheritance

[4] い・う

遺伝率 [力]	heritability	陰　葉	shade leaf
遺伝力 [率]	heritability		
移動追込柵	drift fence		**う**
移動ケージ	movable cage		
移動牧柵	movable fence	ウィスキーかす【粕】	whisky by-product feed
稲　作	rice cultivation [culture]	ウイルス	virus
稲わら【藁】	rice straw	ウイルス病	virus disease
イネ科	Gramineae	ウインチ	winch
イネ科飼料作物	gramineous forage crop	ウインドロー, 集草列, 地干し列	windrow
イネ科 (牧) 草	grass, forage grass		
イネ科牧草地	grass pasture	植付け, 定植, 栽植	planting
移牧《農業類型》	transhumance		
		植付 [栽植] 間隔	plant spacing, planting distance
いや【忌】地	sick soil, soil sickness		
入会権	rights of common	ウエファー成形機	waferer
入会地, 共有地	common [communal] land		
		植え溝	planting furrow
入替え, 置換	replacement	雨　害	rain damage
インキュベーション, 定温培養	incubation	雨季, 梅雨	rainy season
		牛	cattle
		牛 (の)	bovine
因子, 要因	factor	薄播き	sparse seeding [sowing]
因子分析	factor analysis		
陰　樹	shade tree [beares]	うっ閉《林業》	canopy [crown] closure
飲水施設	guzzler, drinker, watering point		
		うっ閉度	crown density
インドール酢酸	indole acetic acid, IAA	うね【畝, 畦】	hill, ridge
		うね【畦】立て	ridge making
インビトロ, 試験管内の, 生体外の	in vitro	うね【畦】立て栽培	ridge culture
		うね【畦】幅	ridge breadth
インビトロ培養, 試験管内培養	in vitro culture	うね【畦】間	row space
		馬	horse
インヒビター, 阻害剤	inhibitor	羽毛粉, フェザーミール	feather meal
インビボ, 生体内の	in vivo	裏　作	secondary [winter] cropping
インブレッドライン, 近交系 (統)	inbred line	雨量計	rain gauge
		ウレアーゼ	urease

え

えい【穎】, 包えい【穎】	glume
えい【穎】花, えい【穎】果	caryopsis
えい【穎】果, えい【穎】花	caryopsis
永久しおれ	permanent wilting
永続性	persistency, durability
永年生［越冬］器官	perennation organ
永年（牧）草地	permanent grassland
営農体系	farming system
栄養	nutrition
栄養価	nutritive value
栄養器官	vegetative organ
栄養系, クローン	clone
栄養系選抜	clonal selection
栄養系繁殖, クローン繁殖	clonal propagation
栄養系分離	clonal separation
栄養失調（症）, 栄養不良	malnutrition
栄養［養分］ストレス	nutrient stress
栄養成［生］長	vegetative growth
栄養成［生］長期	vegetative stage
栄養成分	nutritive component
栄養素, 栄養分	nutrient
栄養素欠乏	nutrient deficiency
栄養素利用効率	nutrient-use efficiency
栄養（体）生殖［繁殖］	vegetative reproduction
栄養（体）繁殖［生殖］	vegetative reproduction
栄養比	nutritive ratio
栄養評価	evaluation of nutritive value
栄養不足	undernutrition
栄養不足の［不良の］, 貧栄養の	oligotrophic
栄養不良, 栄養失調（症）	malnutrition
栄養不良の［不足の］, 貧栄養の	oligotrophic
栄養分, 栄養素	nutrient
エーテル抽出物	ether extract
腋芽	axillary bud
液状きゅう【厩】肥	slurry barnyard manure
液相	liquid phase
液体窒素	liquid nitrogen
液状肥料	fluid fertilizer
液（体）肥（料）	liquid fertilizer [manure]
液肥散布機	liquid manure distributor
液胞	vacuole
エコタイプ, 生態型	ecotype, biotype
エコツーリズム	ecotourism
餌, 飼料	feed (stuff), diet, fodder
餌選択, 飼料選択	diet selection
え【壊】死, ネクローシス	necrosis
S字型成［生］長曲線	sigmoid growth curve
エストロジェン	estrogen
えそ【壊疽】［え【壊】死］病はん【斑】	necrotic lesion
枝肉	dressed carcass
枝肉歩合	dressing percentage
エタノール, エチルアルコール	ethanol, ethyl alcohol

[6] え・お

エチルアルコール, エタノール	ethyl alcohol, ethanol
エチレン	ethylene
越夏性	summer survival
餌づけ飼料	creep feed
越冬	overwintering, wintering
越冬芽	overwintering bud
越冬[永年生]器官	perennation organ
越冬性	winter survival
越年草	winter annual
エネルギー効率	energy efficiency
エネルギー消費	energy expenditure, energy consumption
エネルギー出納	energy balance
エネルギー代謝	energy metabolism
エネルギー転換率	energy conversion efficiency (rate), energy transfer efficiency, energy exchange rate
エネルギー要求	energy requirement
F検定	F-test
F分布	F-distribution
F_1雑種, 一代雑種	F_1 hybrid
エミッション, 放出	emission
えん【嚥】下	swallowing
煙害	smoke injury
塩化ナトリウム, 食塩	sodium chloride, salt
塩基	base
塩基交換容量	cation [base] exchange capacity, CEC
塩基性アミノ酸	basic amino acid
塩基性肥料	basic fertilizer
塩基飽和度	base saturation
	percentage
炎光分光分析	flame spectrophotometry
エンザイム, 酵素	enzyme
塩酸	hydrochloric acid
エンシレージ	ensilage
円すい【錐】花序, 穂	panicle
塩(性, 類)化	salinization
塩生植物	halophyte
塩(性)水, 生理食塩水	saline water
塩素	chlorine, Cl
延長, 伸長	elongation, extension
延展器	spreader
エンドトキシン, 内毒素, 菌体内毒素	endotoxin
エンドファイト, 内生菌	endophyte
塩分	salinity
塩類	salts
塩類集積	salt accumulation
塩類障害, 濃度障害	salt damage [injury]
塩類土壌	saline soil

お

追込牛舎	open cattle shed
追込式畜舎	pen barn
追込場	paddock, stock yard, yard
追播き, 追播	over-seeding [sowing], reseeding
横が【臥】	lying
黄熟期	yellow-ripe stage
黄白化	etiolation
黄変, 白化, クロロシス	chlorosis

応用生態学	applied ecology
大型機械化体系	large-size power farming system
大型農機具	large-size farm machinery
大型分生子, 大型分生胞子	macroconidium [*pl.* macroconidia]
大型分生胞子, 大型分生子	macroconidium [*pl.* macroconidia]
オーキシン	auxin
オートラジオグラフ	autoradiograph
屋外周年飼養	yearlong outdoor feeding
晩生（の）	late maturing
押しつぶし法	squash method
雄しべ, 雄ずい	stamen
汚　水	sewage, polluted water
雄牛, 種牛	bull
雄　花	male [staminate] flower
雄めん【緬】羊	ram
汚　染	pollution
汚染土壌	contaminated soil
汚染物質	pollutant
晩刈り	late cutting
遅播き, 晩播	late seeding [sowing]
オゾン	ozone
オゾン層	ozone layer
途中段階《遷移の》	seral stage
汚泥, スラッジ	sludge
汚泥肥料	sewage sludge fertilizer
帯状耕うん【耘】	band tilling [plowing]
帯状栽培	alley [stripe] cropping
帯状播種	band seeding [sowing]
帯状［成帯］分布, ゾーネーション	zonation
オフセットハロー	offset harrow
親子相関	parent-offspring correlation
オリゴ糖, 少糖	oligosaccharide
オルガネラ, 細胞（小）器官	organella
温室, ガラス室	greenhouse
温室効果	greenhouse effect
温室効果ガス	greenhouse gases
温浸, 消化	digestion
温　帯	temperate zone
温　度	temperature
温度計	thermometer
温度反応	temperature response

か

科	family
外因性の, 外生の	exogenous
外えい【穎】	lemma
開　花	anthesis, blooming, flowering
開花期	blooming stage, flowering time
開花習性	flowering habit
開花日	flowering date
外果皮	epicarp, exocarp
開花誘起	flowering induction
開花要因	flowering factor
海岸植生	coast vegetation
回　帰	regression
回帰曲線	regression curve
回帰係数	regression coefficient
回帰直線	regression line
回帰分析	regression analysis
塊　茎	tuber
塊　根	root tuber
開墾, 開拓	reclamation

[8]　か

開墾地, 開拓地	reclaimed land	改良草地, 改良放牧地	improved pasture
カイ自乗検定	chi-square test	改良品種	improved variety
回収(率), 回復, 歩留り	recovery	改良放牧地, 改良草地	improved pasture
回数, 頻度	frequency		
外生の, 外因性の	exogenous	カウデー	cow-day, CD
		化学処理	chemical treatment
階層, 層	stratum	化学成分	chemical component
階層構造, 層別化	stratification	化学組成	chemical composition
		化学的酸素要求量	chemical oxygen demand, COD
開拓, 開墾	reclamation		
開拓地, 開墾地	reclaimed land	化学的防除, 薬剤防除	chemical control
害虫	insect pest, pest insect		
害虫防除	pest control	化学添加剤	chemical additive
快適性, アメニティ	amenity	化学肥料	chemical fertilizer
		花芽分化	flower bud differentiation, floral differentiation
回転, ターンオーバー	turnover		
解凍	defrosting	花管	corolla tube
解糖	glycolysis	過感受性, 過敏性, 過敏症	hypersensitivity
カイネチン, キネチン	kinetin		
外はい【胚】葉	ectoderm	河岸草地	riparian grassland
開発, 発達	development	花器伝染	flower infection
皆伐《林業》	clear cutting	夏季放牧	summer (season) grazing
開発計画	development planning		
外被, か皮, 殻皮	crust	寡菌動物	specific pathogen free, SPF
回復, 回収(率), 歩留り	recovery	核	nucleus
		がく	calyx, vase
外部形態	external morphology, external feature	核型	karyotype
		拡散	diffusion
灰分	ash	核酸	nucleic acid
解剖学	anatomy	拡散抵抗	diffusion resistance
解剖学的	anatomical	学習	learning
開放受[授]粉, 放任受[授]粉	open pollination	獲得形質	acquired character
		殻皮, 外被, か皮	crust
開やく【葯】	anthesis		
外来遺伝子	foreign gene	角皮貫入	cuticular penetration
外来雑草	exotic weed	核膜	nuclear membrane, septum
外来種	exotic species		
改良	improvement		

核様体, ピレノイド	pyrenoid	加水分解	hydrolysis
		加水分解酵素	hydrolase
かく【撹】乱	disturbance	ガスクロマトグラフィ	gas chromatography
かく【撹】乱地	disturbed habitat [land]	ガス交換	gas exchange
隔　離	isolation	風	wind
隔離栽培	isolated culture	花成素, フロリゲン	florigen
隔離ほ【圃】場	isolation field		
家系, 系統群	family	化成肥料	compound fertilizer
囲　い	enclosure	花成ホルモン	flowering hormone
花こう【梗】, 花柄	peduncle	化石燃料	fossil fuels
		風散布	wind dispersal
加　工	processing	仮説, 仮定	hypothesis
花こう【崗】岩	granite	河川敷	river terrace
か【禾】穀類, 穀物	cereal	下　層	lower layer
		画像解析	image analysis
重ねワク法	nest quadrat method	下層採食	bottom grazing
火山灰	volcanic ash	下層植生, 下草, 下繁草	bottom grass, undergrowth, understory
火山灰土壌	volcanic ash soil		
可視光線	visible ray		
加　湿	humidification	画像処理	image processing
花　序	inflorescence	下層〔深層〕施肥	deep placement, deep application of fertilizer
可消化エネルギー	digestible energy, DE		
可消化乾物収量	digestible dry matter yield	下層土, 心土	subsoil
		可塑性	plasticity
可消化純タンパク【蛋白】質	digestible true protein	カチオン	cation
		家　畜	farm animal, livestock
可消化成分, 可消化養分	digestible nutrient	家畜化, 栽培化	domestication
		家畜管理	animal management
可消化粗タンパク【蛋白】質	digestible crude protein, DCP	家畜飼養	animal feeding
		家畜生産	animal production
可消化養分, 可消化成分	digestible nutrient	家畜生理	animal physiology
		家畜単位	animal unit
可消化養分総量	total digestible nutrients, TDN	家畜ふん【糞】尿	animal manure, livestock waste
過剰障害	excess damage	家畜放牧管理	livestock-grazing regime
過剰調節	overcompensation		
可照日数	possible duration of sunshine	褐色森林土	brown forest soil
		活性汚泥	active sludge
花穂, 穂, 穂状花序	spike	活性化, ふ【賦】活（化）	activation

[10] か

活性化因子, ふ【賦】活剤, 活性化物質	activator
活性化物質, ふ【賦】活剤, 活性化因子	activator
活性物質	active substance
カッタバー	cutter bar
カッタブロア	cutter blower
褐変	browning
褐毛和種牛	Japanese Brown cattle
活力	vigor
活力度	vitality
仮定, 仮説	hypothesis
果糖, フラクトース	fructose
仮道管	tracheid
可働日数	workable days
カドミウム	cadmium, Cd
加入	recruitment
加入パターン［様式］	recruitment pattern
加入様式［パターン］	recruitment pattern
加熱乾燥, 熱風乾燥	heated-air drying
加熱処理	heating treatment
下はい【胚】軸, はい【胚】軸	hypocotyl
下繁草, 下草, 下層植生	bottom grass, undergrowth, understory
過繁地, パッチ	patch
過繁茂	over luxuriant [rank] growth
か皮, 外被, 殻皮	crust
かび	mold [mould]
かび臭	moldy smell
かび毒, マイコトキシン	mycotoxin
かび類, 菌類, 真菌類	fungus [pl. fungi]
過敏症, 過敏性, 過感受性	hypersensitivity
過敏性, 過感受性, 過敏症	hypersensitivity
株	stand, plant
株型, そう【叢】状型, そう【叢】生型	bunch [tufted] type
株型の	bunch, tufted, tussock
カプセル	capsule
株間	interrow space
株密度	stock [stubble] density
花粉	pollen
花粉親	pollen parent
花粉管	pollen tube
花粉ねん【稔】性	pollen fertility
花粉ばい【媒】介, 送粉, 受粉	pollination
花粉ばい【媒】介	pollinator
昆虫	
花粉ばい【媒】介者, 送粉者, ポリネータ	pollinator
花粉飛散	pollen dispersal
花粉母細胞	pollen mother cell
花柄, 花こう【梗】	peduncle
可変性, 変異性	variability
芽胞, 胞子	spore
過放牧	overgrazing
か【噛】み, 食いちぎり, バイト	bite
花蜜	nectar
可溶無窒素物	nitrogen free extract, NFE
殻, シェル	shell
ガラクトース	galactose
ガラス室, 温室	greenhouse

か [11]

カラムクロマトグラフィ	column chromatography
ガリー浸食	gully erosion
カリウム	potassium, K
刈株	stubble
刈(株)跡(地)放牧	stubble grazing, stub-pasturage
刈取り	cutting
刈取回数, 刈取頻度	cutting frequency
刈取間隔	cutting interval
刈取管理	cutting management
刈取期	cutting time, date of cutting
刈取機, モーア	mower
刈取基準	fixed cutting management
刈取結束機	binder
刈取収量	cutting yield
刈取ステージ	cutting stage
刈取高さ	cutting height
刈取適期	optimum stage for cutting
刈取頻度, 刈取回数	cutting frequency
刈取法	cutting method
刈残し	remaining crop after harvesting
刈幅	gathering width, swath
刈払い機, ブッシュカッタ	bush cutter
仮比重	apparent specific gravity, bulk density
カリ肥料	potassium fertilizer
火力乾燥	drying by heating
火力乾燥機, 熱風乾燥機	heated-air dryer
過リン酸石灰	superphosphate
過リン酸分解	phosphorylase
酵素, フォスフォリラーゼ	
カルシウム	calcium, Ca
カルシウム沈着, 石灰集積作用	calcification
カルシフェロール, ビタミンD	calciferol, vitamin D
カルス	callus
カルス誘導	callus induction
カルチパッカ	culti-packer
カルチベータ	cultivator
カルビン回路	Calvin cycle
加令[齢], 老化	ageing, senescence
カロチノイド	carotenoid
かん【稈】	culm, stalk
簡易検定法	easy examination method
簡易更新	simple [rough] renovation
簡易草地造成	simple pasture establishment
感温周期性	thermoperiodicity
感温性	sensitivity to temperature
寒害	cold damage [injury]
干[旱]害	drought damage [injury]
かんがい【灌漑】	irrigation
かんがい【灌漑】草地	irrigated grassland
乾季	dry season
環境	environment
環境アセスメント, 環境評価	environmental assessment
環境影響	environmental impact
環境影響評価	environmental impact assessment
環境汚染	environmental pollution
環境順化	environmental acclimatization

[12]　か

環境条件	environmental condition
環境政策	environmental policy
環境調和型農業	environmentally sound agriculture
環境評価，環境アセスメント	environmental assessment
環境保全	environmental conservation
環境モニタリング	environmental monitoring
環境要因	environmental factor
環境リスクアセスメント	environmental risk assessment
環境劣化［悪化］	environmental degradation
還元	reduction
還元糖	reducing sugar
還元［減数，成熟］分裂	meiosis
還元力	reduction ability, reducing power
観光	tourism
感光性	photosensitivity, sensitivity to daylength
緩効性［遅効性］肥料	slow [delayed] release fertilizer
慣行放牧	conventional grazing (system)
冠根	crown [nodal] root
間作	inter-cropping
完熟	full maturity
完熟期	full-ripe stage
完熟たい【堆】肥	full-fermented compost
感受性，感度	susceptibility, sensitivity
干渉	interference
管状花	tubular flower
緩衝作用	buffer action
干渉作用，交差耐性	cross protection
緩衝能	buffering capacity
冠水	overhead flooding, submergence
かん【灌】水	watering
含水量，水分含量	moisture [water] content
乾生植物	xerophyte
間接効果	indirect effect
感染	infection, affection
完全［有性］時代《カビ》	perfect state
完全優勢	complete dominance
乾燥	drying, desiccation, drought
乾草	hay
乾燥した《土地》	arid
乾草揚げ機，ヘイリフト	hay lift
乾草圧砕機，ヘイコンディショナ	hay conditioner
乾草圧縮機，ヘイプレス	hay press, stationary hay baler
乾草運搬機	hay carrier
乾草価	hay value
乾燥機	dryer
乾草規格	hay grade
乾燥気候	arid climate
乾草こん【梱】包	hay baling
乾草こん【梱】包機，ヘイベーラ	hay baler
乾燥剤	desiccant
乾燥施設	drying plant
乾燥飼料	dried feed
乾燥ストレス	drought stress
乾草生産	hay production
乾燥草地	arid grassland
乾燥速度	drying rate

乾燥地	arid land
乾燥地帯	arid zone [region]
乾草調製	hay making
乾草積上機, ヘイスタッカ	hay stacker
乾草積込機, ヘイローダ	hay loader
乾草破砕機, ヘイクラッシャ	hay crusher
乾草反転機, ヘイテッダ	hay tedder
乾燥ビール酵母	brewer's dried yeast
乾草品質	hay quality
寒　帯	arctic zone
干拓, 土地造成	land reclamation
干拓地	polder
間断かんがい【灌漑】	intermittent irrigation
寒地型イネ科牧草	temperate [cool season] grass
寒地型 (の)	temperate
乾地農業	dry land farming
かん【稈】長	culm length
感度, 感受性	susceptibility, sensitivity
乾土効果	soil drying effect
乾　乳	dry off
乾乳牛	dry cow
官能試験	sensory test
干[旱]ばつ【魃】	drought
間伐強度	thinning intensity
干[旱]ばつ【魃】指数	aridity index
干[旱]ばつ【魃】度	aridity
間伐頻度《林業》	thinning frequency
冠　部	crown
乾　物	dry matter
乾 (物) 重	dry (matter) weight
乾物収量	dry matter yield
乾物消化率	dry matter digestibility
乾物消失率	dry matter disappearance rate
乾物生産, 乾物生産量	dry matter production
乾物生産量, 乾物生産	dry matter production
乾物生産力	dry matter productivity
乾物蓄積	dry matter accumulation
乾物分配率	dry matter partitioning ratio
乾物率	dry matter ratio
冠部被度	crown cover
γ線	gamma ray
含有率, 含有量, 含量	content
含有量, 含量, 含有率	content
管　理	management, care
含硫アミノ酸	sulfur amino acid
含量, 含有量, 含有率	content
寒冷紗	cheese cloth, shade cloth
寒冷地	cold area [region, district]

き

キーストーン種	keystone species
気　温	air temperature
機械化	mechanization
機械化体系	power farming system
機械作業	mechanized working
機会的 [ランダム, 不規則] 分布	random distribution
機会的変動	random drift
機械特性	mechanical property

[14] き

帰化雑草	naturalized weed
帰化種	naturalized species
器　官	organ
気管《動物》	trachea
基幹品種	leading variety
基群集	sociation
奇　形	distortion, malformation
危険分散	dispersion of risk
気　孔	stoma [*pl.* stomata]
気孔開度	stomatal aperture
気孔感染	stomatal infection
気候区	climatic province
気候指数	climatic index
気孔蒸散	stomatal transpiration
気候図	climograph
気候生産指数	meteorological index of production
気候帯	climatic zone
既耕地, 耕地	cultivated [arable] land
気孔抵抗	stomatal resistance
気候変化, 気候変動	climate change
気候変動, 気候変化	climate change
気候要因	climatic factor
気候要素	climatic [meteorological] element
気　根	aerial root
ギ【蟻】酸	formic acid
基　質	ground substance, matrix, substrate
希釈液, 希釈剤	diluent, dilute solution
希釈剤, 希釈液	diluent, dilute solution
技術体系	systematized techniques
基　準	standard
気　象	meteorology
気象生産力	meteorological productivity
キシロース	xylose
寄　生	parasitism, parasitization
寄生型, 病原型	pathotype
寄生者[体], 寄生虫, 寄生生物	parasite
寄生生物, 寄生者[体], 寄生虫	parasite
寄生性分化, 生理的分化	physiological specialization
寄生虫, 寄生者[体], 寄生生物	parasite
規制[限定]要因	limiting factor
季節型	seasonal form
季節性	seasonality
季節生産性	seasonal productivity
季節遷移	seasonal succession
季節繁殖	seasonal breeding
季節変異	seasonal variation
季節変動	seasonal variation [change]
季節放牧	seasonal grazing
気　相	gaseous phase
基礎栄養成[生]長	basic vegetative growth
規則性	regularity
規則分布	regular distribution
基礎飼料	basal diet [feed]
基礎代謝	basal metabolism
期待値	expected value, expectation
キチン	chitin
きっ【拮】抗, きっ【拮】抗作用	antagonism
きっ【拮】抗作用, きっ【拮】抗	antagonism

喫　食	pasture intake, grazing		法, 自由選択法
基底被度	basal coverage [cover]	キャリア, 担体, 保毒[菌]者	carrier
基底面積	basal area	キャリブレーション, 目盛補正, 較正	calibration
木戸《放牧場の》	bumper gate		
機動細胞	motor cell		
キネチン, カイネチン	kinetin		
		牛　衣	cattle clothes
機　能	function	きゅう【嗅】感覚, 臭覚	olfaction
木の実	mast, nut		
揮　発	volatilization	休閑地	fallow (land)
揮発性塩基態窒素	volatile basic nitrogen, VBN	休閑（の）	fallow
		球　茎	corm
揮発性脂肪酸	volatile fatty acid, VFA	球茎《チモシー》	haprocorm
		球形ウイルス	spherical virus
揮発性成分	volatile constituent	急傾斜草地	steep sloping pasture
揮発性物質	volatile substance [matter]	急傾斜地	steep slope, steep sloping land
基　肥	basal application of fertilizer, basal dressing	急傾斜畑	steep sloping field
		吸血昆虫	blood-sucking insect
		吸光係数	light extinction coefficient
忌避剤, 忌避物質, 駆散薬	repellent	給餌機	feeder
		休止状態, 休眠, 休眠性	dormancy
忌避物質, 忌避剤, 駆散薬	repellent		
		吸湿性	hygroscopicity
基　部	base	吸　収	absorption, uptake
気密サイロ	airtight silo	吸収器官	sucker
キメラ	chimera	吸収根	absorbing [sucking] root
逆行変性, 退化, 変性	degeneration		
		吸汁昆虫	sucking insect
客　土	soil dressing	牛床, ストール	stall
逆流, 吐戻し, 弁閉鎖不全	regurgitation	吸　水	water absorption
		給水場	watering place
ギャップ, 空げき【隙】	gap	牛スラリー	cattle slurry
		急性の	acute
ギャップの大きさ	gap size	急速高温度乾燥	dehydration
		休息場, たてば	resting place
キャノピ, 草冠, 林冠	canopy	吸　着	adsorption
		牛　道	cattle trail [track]
キャフェテリア	cafeteria method	牛　乳	milk

[16] き

牛　尿	cattle urine	共生的窒素固定	symbiotic nitrogen fixation
きゅう【厩】肥，畜肥，肥［こえ］	stable (barnyard) manure, manure	競争排除	competitive exclusion
		競争密度効果	competition-density effect, C-D effect
きゅう【厩】肥車	dung cart	競争力	competitive ability
キューブ	cube	共　存	coexistence, cooccurrence
牛ふん【糞】	cattle feces		
牛ふん【糞】たい【堆】肥，牛ふん【糞】尿	cattle manure	共存種	binding species
		協同，協同作業	cooperation
		協同作業，協同	cooperation
牛ふん【糞】尿，牛ふん【糞】たい【堆】肥	cattle manure	共同社会，共同体	community
		共同体，共同社会	community
休　牧	rest, non-grazing	共同放牧地	common [communal] pasture
休牧期間	non-grazing [rest] period		
		共同利用模範牧場	pilot livestock ranch for common use, pilot communal livestock farm
休眠，休眠性，休止状態	dormancy		
休眠芽	dormant [latent] bud		
休眠覚醒	dormancy awakening		
休眠型	dormancy form	共分散	covariance
休眠期	dormant stage, resting period	共分散分析	analysis of covariance
		共有地，入会地	common [communal] land
休眠期間	dormant period		
休眠種子	dormant seed	供与体，ドナー	donor
休眠性，休眠，休止状態	dormancy	行列モデル	matrix model
		極期，極相，クライマックス	climax
休眠打破	dormancy breaking		
給　与	feeding		
凝灰岩	tuff	局所性，地域性	locality
供給草量，割り当て草量	forage [herbage, pasture] allowance	極相，極期，クライマックス	climax
競合［争］，競合い	competition		
		極相群落，クライマックス群集［植生］	climax community [vegetation]
胸高直径	diameter at breast height, DBH		
凝集，集合	aggregation		
共進化	coevolution	極相群落優占種，クライマックス優占種	climax dominants
共生，相利共生	symbiosis, mutualism		
強制休眠	enforced dormancy	極相林	climax forest

局地気候	local climate	近赤外光	far-red light
局地処理	spot treatment	近赤外分光法	near-infrared reflectance spectroscopy
局部病はん【斑】	local lesion		
去勢《動物》, 除雄《植物》, 性腺摘除《動物》	castration, emasculation	菌そう【叢】, コロニー, 群体	colony
去勢雄牛《性成熟後》	ox	金属（結合）タンパク【蛋白】質	metalloprotein
去勢牛	steer	菌体内毒素, 内毒素, エンドトキシン	endotoxin
挙動, 行動	behavior		
魚粉	fish meal		
切穂	ear breaking	禁牧	prohibition of grazing, livestock exclusion
キレート	chelate		
均一性	evenness		
近縁種	allied [related] species	禁牧草地	closed pasture
		近隣個体	neighbor
近縁野生種	wild relative	菌類, かび類, 真菌類	fungus [pl. fungi]
菌核	sclerotium [pl. sclerotia]		
		菌類病	fungal [fungous] disease
菌系, レース	race		
近交係数	inbreeding coefficient		
近交系（統）, インブレッドライン	inbred line	**く**	
		食いちぎり, か【噛】み, バイト	bite
近交弱勢［退化］	inbreeding depression		
近交退化［弱勢］	inbreeding depression		
近郊農業	suburban agriculture	空間的異質性, 空間不均一性	spatial heterogeneity
菌根	mycorrhiza		
菌根菌	arbuscular mycorrhizae	空間不均一性, 空間的異質性	spatial heterogeneity
菌糸, 菌糸体	mycelium [pl. mycelia], hypha [pl. hyphae]	空間分布	spatial distribution
		空間変異	spatial variation
		空気伝染病	air-borne disease
菌糸体, 菌糸	mycelium [pl. mycelia] hypha [pl. hyphae]	空げき【隙】, ギャップ	gap
		空げき【隙】率, 孔げき【隙】率	porosity
均質化	homogenization		
均質度係数	coefficient of homogeneity, CH		
		偶存種	accidental species
近親交配［近交］, 同系交配	inbreeding	空中散布	aerial application
		空中分げつ	aerial tiller(ing)

[18]　く

空腹時，絶食	fasting
クエン酸回路，トリカルボン酸回路	tricarboxylic acid [TCA] cycle
区画法，方形区法，枠法	quadrat method
茎	stem, stalk, haulm
茎挿し	stem cutting
茎の［茎葉，地上部］成［生］長	shoot growth
茎（部），若枝，苗条	shoot
草《茎葉》	herbage
草型	plant type
草刈り	mowing
草・草食獣の共存	grass-grazer mutualism
草丈	plant length
草床形成	sod formation
草床をつくる草	sod grass
駆散薬，忌避剤，忌避物質	repellent
駆除，防除	control
クチクラ	cuticle
クチクラ層	cuticular layer
屈光性	phototropism
屈日性	heliotropism
屈地現象，屈地性，向地性	geotropism
屈地性，屈地現象，向地性	geotropism
駆動トレーラ	power-driven trailer
駆動輪	driving wheel
苦土石灰	magnesium-calcium carbonate, ground dolomitic limestone
クノップ液	Knop's solution
クマリン	coumarin
組合せ能力	combining ability
組換価	recombination value
組換型	recombinant
グライ層	gley horizon
クライマックス，極相，極期	climax
クライマックス群集［植生］，極相群落	climax community [vegetation]
クライマックス植生［群集］，極相群落	climax community [vegetation]
クライマックス優占種，極群落優占種	climax dominants
グラスサイレージ	grass silage
クラスター分析	cluster analysis
グラステタニー	grass tetany
クランプサイロ	clamp silo
クリープ放牧	creep grazing
グリシンベタイン	glycine betaine
グルコアミラーゼ	glucoamylase
グルコース，ブドウ糖	glucose
グルコシド，配糖体	glucoside
グレーンドリル	grain drill
クレブス回路	Krebs cycle
グロースキャビネット，植物育成箱	growth cabinet
クローナル成［生］長，クローン成［生］長	clonal growth
クローン，栄養系	clone
クローン化	cloning
クローン構造	clone structure
クローン成［生］長，クローナル成［生］長	clonal growth

クローン繁殖, 栄養系繁殖	clonal propagation		**け**
黒毛和種牛	Japanese Black cattle		
黒泥土	muck soil	経営規模	size of farm
グロブリン	globulin	経営計算	business [farm] accounting
黒ボク土	Andosol		
クロマトフォア, 有色体, 色素体	chromatophore	経営収支	management [business] balance
		経営組織	business [farm] organization
クロモーゲン, 色素原	chromogen	経営分析	analysis of farm operation
クロモソーム, 染色体	chromosome	けい【珪】化	silicification
クロロシス, 黄変, 白化	chlorosis	景観, ランドスケープ	landscape
クロロフィル, 葉緑素	chlorophyll	景観シミュレーション	landscape simulation
群《動物》	herd	景観設計	landscape design
群管理	herd management	傾向, 動向	trend
群給餌	group feeding	蛍光	fluorescence
群系	formation	蛍光X線分析	fluorescent X-ray analysis
群集, 群落	community, association		
		経口摂取, 摂取	ingestion
群集生態学	community ecology	経済[実用, 農業]形質	economic [agronomic] character [trait]
群集動態(学)	community dynamics		
くん【燻】蒸剤	fumigant		
群体, 菌そう【叢】, コロニー	colony	経済的な	economical
		経済の	economic
		ケイ酸	silicic acid
群団	alliance	形質, 特性	character
群度	sociability	形質転換	transformation
群の大きさ	herd size	形質発現	phenotypic expression, manifestation
群落, 群集	community, association		
		傾斜, 斜面	slope
群落光合成	photosynthesis of a community	傾斜(角)度	inclination, slope [slant] angle
群落構造	community structure	傾斜採草地	sloping [hillside] meadow
群(落)生態学	synecology		
群落測度	community measures	傾斜草地	sloping pasture, hillside grassland

[20] け

傾斜地	slope land, sloping land, slope
傾斜 [斜面] 方位	slope aspect
茎 重	stem [tiller] weight
軽しょく【埴】土	light clay
系図, 血統	pedigree
茎 数	number of tillers [stems]
形成層	cambium
ケイ素	silicon, Si
計測装置	measuring instrument
形態 (学)	morphology
形態形成 [発生]	morphogenesis
形態発生 [形成]	morphogenesis
携帯用草量計	portable grass meter, portable pasture meter
茎 頂	shoot apex
系 統	strain
系統育種法	pedigree breeding [method]
系統 (間) 変異	phylogenetic variation
系統群, 家系	family
系統選抜	line [pedigree] selection
系統的抽出	systematic sampling
系統的配置	systematic arrangement
軽度放牧	light [lenient] grazing
経年草地	ageing pasture [meadow], aged [old] grassland
繋牧, 繋飼い	tether grazing, tethering
けい【畦】面被覆, 敷草, マルチング	mulching
茎葉, 葉《集合的》	foliage
茎葉飼料	forage
茎葉 [地上部, 茎の] 成 [生] 長	shoot growth
茎葉比	stem/leaf ratio
経卵伝染	transovarial passage
ケージ	cage
下水汚泥	sewage sludge
血 液	blood
欠 株	lost [missing] plant
ケツ【頁】岩	shale
結合剤	binder
血色素, ヘモグロビン	hemoglobin
結 実	fructification, seed setting
結実率	seed set percentage
げっ歯類, 野ねずみ	rodent
欠測値	missing plot [value]
血統, 系図	pedigree
血 糖	blood glucose
欠 乏	deficiency
欠乏症 (状)	deficiency symptom
月令 [齢]	age in month
ケトーシス, ケトン症	ketosis
解 毒	detoxication
ケトン症, ケトーシス	ketosis
ゲノム	genome
ゲノム分析	genome analysis
下 痢	diarrhea
ゲ ル	gel
ゲル化	gelation
ケルダール法	Kjeldahl method
検えき【疫】, 防えき【疫】	quarantine
限界温度, 臨界温度	critical temperature
限界日長	critical day length
限外濾過	ultrafiltration

減価償却	depreciation	減退, 衰弱, 低下	decline
原 基	primordium [*pl.* primordia]	原虫, 原生動物, プロトゾア	protozoa
嫌気呼吸, 無気呼吸	anaerobic respiration	検 定	test
嫌気(性)細菌	anaerobic bacteria [microbe]	限定 [規制] 要因	limiting factor
嫌気的条件	anaerobic condition	顕 熱	sensible heat
嫌気的分解	anaerobic degradation	玄武岩	basalt
兼業農家	part-time farm household, sidework farmer	ケンブリッジローラ	Cambridge roller
		研磨, 粉砕	grinding
原形質	protoplasm	兼用種《家畜》	dual purpose breed
原形質体, プロトプラスト	protoplast	検量線	calibration equation

こ

原形質膜	plasma membrane		
原々種	foundation seed	高位安定生産	high and stable production
嫌光性種子	light-inhibited seed	高位生産性	high productivity
検査, 検診	inspection	降 雨	rainfall
原産地	place of origin, origin	降雨強度	rainfall intensity
原子吸光法	atomic absorption spectrochemical analysis	降雨侵食	rainfall erosion
		降雨装置	rainfall apparatus
絹糸抽出期	silking stage	耕うん【耘】	plowing and harrowing, tillage, cultivation
原種, もと種《作物》	original [stock] seed		
原種《家畜》	original breed	耕うん【耘】機	tiller
減 収	yield decrease	恒温, 定温	constant [fixed] temperature
原種ほ【圃】	stock seed field		
検診, 検査	inspection	高温障害	high temperature injury
減数 [成熟, 還元] 分裂	meiosis	高温ストレス	high temperature stress
原生動物, 原虫, プロトゾア	protozoa	硬化, ハードニング	hardening
原生林	virgin forest	光化学反応	photochemical reaction
嫌石灰植物	calciphobous plant		
元素, 要素	element	抗かび剤	mold inhibitor
現存量	standing crop, biomass	交換酸度	exchangeable acidity
		交換性塩基, 置換性塩基	exchangeable base
検体, 標本, 試料	sample		

[22] こ

耕起	plowing
後期合成世代	advanced synthetic generation
耕起深度〔耕深〕	depth of plowing [tillage]
好気性（細）菌	aerobic bacteria
好気性生物	aerobe
香気成分	aroma component
後期世代	advanced generation
好気的変敗	aerobic deterioration
耕起法	plowing [tilling] method
公共育成牧場	public pasture for raising cattle
公共草地	cooperative [public] grassland
公共牧場	cooperative [public] pasture, cooperative [public] stock farm
抗菌性の，抗生物質	antibiotic
航空機散布	airplane spraying
航空播種	aerial seeding [sowing]
孔げき【隙】率，空げき【隙】率	porosity
抗血清	antiserum [*pl.* antisera]
抗原	antigen
高原草地	hilly grassland
光合成	photosynthesis
光合成器官	photosynthetic organ
光合成効率	photosynthetic efficiency
光合成産物	photosynthate, photosynthetic product
好光性種子	heliophilous [light-favored] seed, light germinating seed
光合成速度	photosynthetic rate
光合成組織	photosynthetic tissue
光合成能（力）	photosynthetic capacity
交互作用（項），相互作用	interaction
交互条播	alternate row seeding [sowing]
交叉，乗換え，交差	crossing over
交差，交叉，乗換え	crossing over
鉱さい，スラグ	slag
耕作，栽培	culture, cultivation
交差耐性，干渉作用	cross protection
交雑，雑種形成	crossing, hybridization
交雑育種	cross [hybridization] breeding
交雑種，雑種	crossbred, hybrid
交雑ねん【稔】性，交雑率	crossing rate
交雑不和合性，交配不和合性	cross incompatibility
交雑率，交雑ねん【稔】性	crossing rate
抗酸化剤，抗酸化の	antioxidant
抗酸化の，抗酸化剤	antioxidant
高山採草地	alpine meadow
高山草地	alpine grassland
高山帯	alpine zone
高山地方	alpine region
高山の	alpine
子牛	calf, veal calf
硬実	hard seed
鉱質土壌	mineral soil
格子配列法	lattice design

高次 [位] 分げつ	high order tiller, higher-nodal-position tiller
耕 種	field husbandry
光周期	photoperiodism, light rhythm
光周 [日長] 反応	photoperiodic response
後 熟	after-ripening
耕種的防除	cultural control
甲状腺	thyroid gland
子牛用離乳飼料	calf starter
更 新	renovation, regeneration
更新草地	renovated pasture [grassland]
更新伐	regeneration cutting [felling]
更新用家畜	replacement stock
洪水, 湛水	flooding, waterlogging, submergence
高水分サイレージ	high moisture silage
降水 (量)	(amount of) precipitation
較正, 目盛補正, キャリブレーション	calibration
合 成	synthesis
合成第1代	first synthetic generation, syn1
合成品種	synthetic variety
抗生物質, 抗菌性の	antibiotic
洪積台地	diluvial plateau
洪積地 (層)	diluvium
洪積土	diluvial soil
抗線虫薬, 殺線虫剤	nematocide
酵素, エンザイム	enzyme
構 造	structure
構造性炭水化物	structural carbohydrates
酵素反応	enzyme reaction
恒存種	constant species
恒存度, 常在度	constance
抗 体	antibody
後代, 子孫, 次代	progeny, offspring
後代検定	progeny test
耕地, 既耕地	cultivated [arable] land
向地性, 屈地現象, 屈地性	geotropism
高地草地	upland grassland
口てい【蹄】えき【疫】	foot and mouth disease
好適範囲	optimum range
耕 土	arable soil
硬 度	hardness
行動, 挙動	behavior
行動型	behavior pattern
行動生態学	behavioral ecology
高度化成肥料	high-analysis mixed fertilizer
購入価格	purchased price
購入飼料	purchased feed
荒廃, 劣化, 衰退	deterioration, degradation
交配, 接合, 交尾	mating
こう【勾】配	gradient
交配 [接合] 型	mating type
交配組合せ	cross combination
荒廃 [衰退] 草地	degrading grassland, waste grassland, deteriolated grassland
交配不和合性, 交雑不和合性	cross incompatibility
耕 盤	plow sole

広範な, 大規模な	extensive
交尾, 交配, 接合	mating
高泌乳牛	high producing dairy cow
厚壁胞子, 厚膜胞子	chlamydospore
酵母, イースト	yeast
高木層	tree stratum
厚膜胞子, 厚壁胞子	chlamydospore
公有林	communal [community] forest
広葉雑草	broad-leaved weed
広葉樹	broad-leaved tree
広葉樹林	broad-leaved forest
広葉草本	forb
効　率	efficiency
高冷地	high altitude cool-climate region
肥［こえ］, 畜肥, きゅう【厩】肥	stable (barnyard) manure, manure
固液分離機	slurry separator
こえ［肥］焼け	fertilizer injury
コーティング, 被覆	coating
コート種子	coating [coated] seed
コーンハーベスタ, トウモロコシ収穫機	corn harvester
コーンプランタ, トウモロコシ点ま【蒔】き機	corn planter
小型分生子, 小型分生胞子	microconidimu [pl. microconidia]
小型分生胞子, 小型分生子	microconidimu [pl. microconidia]
呼吸基質	respiratory substance
呼吸係数	respiratory quotient, RQ
呼吸酵素	respiratory enzyme
呼吸（作用）	respiration
呼吸能	respiration activity
穀実, 子実	grain
穀草式輪作	ley farming
極晩生（の）	extremely late maturing
穀物, か【禾】穀類	cereal
国有林	national forest
国立公園	national park
国立保護区	national reserve
穀　類	cereal grains
極早生（の）	extremely early maturing
固形飼料, ペレット	pellet
固形分補正乳	solid corrected milk, SCM
誤　差	error
誤差分散	error variance
枯　死	death
枯死（の）	dying
枯死部	dead material [matter]
こ【糊】熟期	dough stage, dough-ripe stage
コスト	cost
互生《葉》	alternate
個生態学	autoecology, ideoecology
固　相	solid phase
個　体	individual
個体植, 1本植	spaced [single] planting
個体群, 集団, 母集団	population
個体群［集団］構造	population structure

個体群成長速度 [生]	crop growth rate, CGR
個体群統計学, デモグラフィ, 人口学	demography
個体群動態	population dynamics
個体群 [数] 密度, 生 [棲] 息密度	population density
個体サイズ《植物》	plant size
個体識別	individual identification
個体重《植物》	plant weight
個体数	number of individuals, population size
個体数過剰	over-population
個体数 [群] 密度, 生 [棲] 息密度	population density
個体選抜	individual selection
個体 [植物体] の成績	plant performance
個体変異	individual variation
個体密度《生物》	individual density
個体密度《植物》	plant density
五炭糖	pentose
鼓脹 [腸] 症, 泡まつ【沫】性鼓脹 [腸] 症	bloat
子付 (繁殖) 雌牛, 授乳牛	suckler cow
固定 (化), 固定法	fixation
固定コドラート	permanent quadrat
固定資産	fixed assets
固定資本	fixed capital
固定窒素	fixed nitrogen
固定トランセクト	permanent transect
固定費用	fixed cost
固定法, 固定 (化)	fixation
固定放牧, 定置放牧	fixed grazing [stocking], set stocking
固定牧柵	fixed fence
コドラート, 方形コドラート, 方形区	quadrat
コバルト	cobalt, Co
こ【糊】粉層, アリューロン層	aleurone layer
米ぬか	rice bran
子めん【緬】羊生産	lamb production
固有種, 在来種	endemic species
個葉	single leaf
互用プラウ	two-way plow
コルク形成	cork formation
コルヒチン	colchicine
ゴルフ場	golf course
コレステロール	cholesterol
コロニー, 菌そう【叢】, 群体	colony
根圧	root pressure
根域	root zone
根冠	root cap
根群, 根系	root system
根系, 根群	root system
根茎, 地下茎	subterranean stem, rhizome
根茎型	rhizomatous type
根茎植物	rhizomatous plant
根圏	rhizosphere
混合乾草	mixed hay
混合 (剤)	mixture
混合サイレージ	mixed silage
混合試料	composite sample

混合飼料	total mixed ration, TMR
混合農業	mixed farming
混（合放）牧	mixed grazing
混在	mixture
根菜類	root crops
混作	mixed [multiple] cropping
コンシステンシー，粘ちょう【稠】度，密度	consistency
根しょう【鞘】	coleorhiza
混植	companion [mixed] planting
混生率［割合］	mixture rate [ratio]
根絶，撲滅	eradication
混層耕プラウ	layer mixing plow
根端	root apex
昆虫	insect
昆虫相	insect fauna
混播，混ぜ播き	mix seeding [sowing]
混播草地	mixed sown sward
混播放牧草地	mixed sown pasture
こん【梱】包，ベール	bale
こん【梱】包乾草	baled hay
こん【梱】包サイレージ	baled silage
混牧林，牧野林	grazing forest, grazing wood land
コンポスト	compost
コンポスト化，たい【堆】肥化，発酵乾燥	composting
根毛	root hair
根粒	root nodule
根粒菌	leguminous [root nodule] bacteria
根粒形成	root nodulation

さ

差異	difference
座位，遺伝子座	locus
催芽，発芽促進	hastening of germination
細菌	bacterium [pl. bacteria]
細菌粘塊	bacterial ooze
細菌病	bacterial disease
細菌漏出	bacterial exudation
最高分げつ期	maximum tiller number stage
細根	rootlet
採餌	foraging
採集	collection
最終収量一定の法則	law of constant final yield
採種（栽培），種子生産	seed production
最上位展開葉	youngest expanding leaf
最小自［二］乗法	method of least squares
最小有意差	least significant difference, LSD
最小律	law of minimum
栽植，植付け，定植	planting
栽植［植付］間隔	plant spacing, planting distance
栽植距離	spacing
採食［摂取］行動	grazing [ingestive] behavior
採食時間	eating time
採食し【嗜】好性，し【嗜】好性	palatability, diet preference
採食草	grazed herbage
栽植方式［様式］	planting pattern

栽植密度	planting density	砕土機,	clod crusher,
栽植様式 [方式]	planting pattern	ディスクハロウ	disk harrow
採食 (量)	intake, herbage intake	採土器	soil sampler
再侵入	recolonization	栽培, 耕作	culture, cultivation
サイズ分布	size distribution	栽培化, 家畜化	domestication
再生, 再成 [生] 長	regrowth	栽培環境	environment for cultivation
再生産	reproduction, regeneration	栽培管理	cultivation, plant husbandry
再生障害	injury of regrowth	栽培技術	cultivation technique
再生草	aftermath	栽培限界	cultivation limit
再生草勢	regrowth vigor	栽培種	cultivated species
再成 [生] 長, 再生	regrowth	栽培条件	cultivation condition
		栽培 [作付] 体系, 作付方式	cropping [cultivation] system
細切	chopping		
細切機	chopper		
採草	mechanical (forage) harvesting, cutting, harvest	栽培品種, 品種	cultivar
		栽培法	cultivation [plant husbandry] method
採草地	meadow		
最大可能成 [生] 長量	potential growth	栽培ほ【圃】場	growing field, cultivated field
最大乾物生産速度	maximum dry matter production rate	栽培様式	cultivation system [pattern]
最大容水量	maximum moisture [water] capacity	再播種	resowing
		細胞	cell
最大葉面積指数	ceiling leaf area index	細胞遺伝学	cytogenetics
最低致死温度	minimum fatal low temperature	細胞外 (の)	extracellular
		細胞学	cytology
最低有効温度	minimum effective temperature	細胞間げき【隙】	intercellular space
		細胞間の	intercellular
最適温度	optimum temperature	細胞間物質	intercellular substance
最適化	optimization	細胞質	cytoplasm
最適クローバ率	optimal clover content	細胞 (小) 器官, オルガネラ	organella
最適採餌	optimal foraging		
最適餌	optimal diet	細胞質遺伝	cytoplasmic inheritance
最適葉面積指数	optimum leaf area index, opt. LAI	細胞質ゲノム	cytoplasmic genome
砕土	clod breaking, soil crushing	細胞質雄性不ねん【稔】	cytoplasmic male sterility
サイトカイニン	cytokinin		

細胞内容（物）	cell(ular) contents, CC
細胞培養	cell culture
細胞分化	cell differentiation
細胞分裂	cell division
細胞壁	cell wall
細胞壁成分	cell wall component [constituent], CWC
細胞壁物質	cell wall substances, CW
細胞膜	cell membrane
細網《組織》, 第二胃	reticulum
在来系統, 在来品種, 在来種	local [native, indigenous] variety
在来種, 固有種	endemic species
在来種, 在来品種, 在来系統	local [native, indigenous] variety
在来（の）, 自生（の）	native, indigenous, spontaneous
在来品種, 在来種, 在来系統	local [native, indigenous] variety
細流浸食	rill erosion
サイレージ	silage
サイレージ調製	ensiling, silage making
サイレージ添加物	silage additive
サイレージ取出し機	silage [silo] unloader
サイレージ発酵	silage fermentation
サイレージ品質	silage quality
サイロ	silo
砂岩	sandstone
砂丘（地）	sand dune [hill]
作業機	working machine
作業効率	working efficiency
作業精度	operating accuracy, quality of work
作業体系	operation [work] system
作業能率	rate of work
柵, 牧柵	fence
作柄	crop situation
作期	cropping season
酢酸	acetic acid
搾汁	dejuicing
作条器	planting opener
柵状組織	palisade tissue
サクセッション, 遷移	succession
作付順序	cropping order [sequence]
作付［栽培］体系	cropping [cultivation] system
作付比率	cropping ratio
作付方式, 栽培［作付］体系	cropping [cultivation] system
作付面積	planted area
作土	cultivation layer
搾乳機, ミルカ	milker
搾乳牛, 泌乳牛	milking [lactating, dairy] cow
搾乳曲線	milking curve
作目	kind of crop
作物	crop
作物遺体, 作物残渣	crop residue
作物残渣, 作物遺体	crop residue
砂耕	sand culture
ササ型草地	Sasa-type grassland, dwarf bamboo pasture
挿木［芽］	vegetative propagation
砂質土壌	sandy soil
砂壌土	sandy loam
雑菌, 腐生者［菌］	saprophyte
殺菌剤, 殺真菌薬	fungicide

さ [29]

雑穀	millet	酸化酵素	oxidase
雑種，交雑種	crossbred, hybrid	酸化的リン酸化（反応）	oxidative phosphorylation
雑種強勢，ヘテローシス	heterosis, hybrid vigor	山間草地	mountain grassland
雑種形成，交雑	crossing, hybridization	山間地帯，山岳地帯，山地	mountainous area
雑種第一代	first filial generation, F_1	散形花序	umbel
		三系交配	three-way cross
殺真菌薬，殺菌剤	fungicide	残根	root residue, residual root
殺線虫剤，抗線虫薬	nematocide	残さ【渣】，残留物	residue
雑草	weed	三次膜《細胞膜の》	tertiary wall
殺草性，植物毒性	phytotoxicity	三出複葉	ternate compound leaf
雑草防除	weed control	酸性，酸度，酸性度	acidity
殺そ剤	rodenticide		
殺だに剤	acaricide	酸性雨	acid rain
殺虫剤	insecticide	酸性化	acidification
砂土	sand	産生株，生産者	producer
里山	satoyama	酸性植物	acid plant
さなぎ【蛹】	pupa	酸性デタージェント繊維	acid detergent fiber, ADF
砂漠	desert		
砂漠化	desertification	酸性デタージェント不溶性窒素	acid detergent insoluble nitrogen, ADIN
砂漠かん【灌】木	desert shrub		
砂漠緑化	desert revegetation [planting]	酸性度，酸性，酸度	acidity
サバンナ《熱帯の草原》	savannah	酸性土壌	acid soil
		酸性肥料	acid manure, acidic fertilizer
サブクライマックス，亜極相	subclimax	酸素	oxygen, O
		残草（量）	post-grazing pasture [herbage] mass, residual herbage (mass)
サポニン	saponin		
さや【莢】	pod, shell		
作用スペクトル	action spectrum		
酸	acid	酸素吸収	oxygen uptake
酸化	oxidation	酸素放出	oxygen evolution
酸化還元電位，Eh	oxidation-reduction [redox] potential	山村	mountain village
		山地，山間地帯，山岳地帯	mountainous area
山岳地帯，山間地帯，山地	mountainous area	山地草原	upland meadow

山地放牧	mountain [alpine] grazing	紫外吸光度法	ultraviolet absorption method
山地放牧地	alpine pasture	紫外線	ultraviolet ray
山地酪農	mountain dairy	紫外線照射	ultraviolet radiation
酸度, 酸性, 酸性度	acidity	自家受精, 自殖	self-fertilization, selfing, autogamy
産乳	milk production	自家受粉	self-pollination
散播	broadcast(ing)	自家受粉[自殖性]作物	self-pollinated crop
散播機	broadcaster		
散布, 分散	dispersion, dispersal		
散布器, 噴霧機	sprayer	自家配合	self-formulation
散布器官型	disseminule form	自家不ねん【稔】性	self-sterility
散布むら	unevenness of application	自家不和合性	self-incompatibility
酸不溶性灰分	acid-insoluble ash	自家和合性	self-compatibility
散粉機	duster	師[篩]管	sieve tube
残葉	foliar residue, residual leaf	時間差放牧	sequential grazing
残留物, 残さ【渣】	residue	時間制限放牧	on-and-off [rationed] grazing
		歯かんハロー	spike tooth harrow
		自記温度計	recording thermometer, thermograph

し

シアン化物生成	cyanogenesis	敷草, けい【畦】面被覆, マルチング	mulching
C-N 比, 炭素窒素比	C-N ratio, carbon-nitrogen ratio		
C-F 比	C-F ratio	色素	pigment
CO_2 上昇	elevated CO_2, CO_2 elevation	色素原, クロモーゲン	chromogen
CO_2 施肥	CO_2 enrichment	色素体, 有色体, クロマトフォア	chromatophore
CO_2 濃度増加	increasing CO_2 concentration	色素体, プラスチド	plastid
C_3 経路	C_3 pathway	色素タンパク【蛋白】質	chromoprotein
C_3 植物	C_3 plant		
シードペレット	seed pellet		
C_4 経路	C_4 pathway	色調	color tone
C_4-ジカルボン酸回路	C_4 dicarboxylic acid cycle	識別, 同定	identification
C_4 植物	C_4 plant	識別種	differential species
ジェネット	genet	自給飼料	self-supplying feed
シェル, 殻	shell	自給肥料	self-supplying manure
しおれ	wilting		

敷料，敷わら【藁】	bedding, litter
敷わら【藁】，敷料	bedding, litter
時空的な	spatiotemporal
資源	resource
試験管内の，インビトロ，生体外の	in vitro
試験管内培養，インビトロ培養	in vitro culture
試験区，実験区	experimental plot
試験設計，実験計画	experimental design, design of experiment
資源配分，資源分配	resource allocation
資源分割	resource partitioning
資源分配，資源分配	resource allocation
試験ほ【圃】	test field
し【嗜】好性，採食し【嗜】好性	palatability, diet preference
地ごしらえ	ground clearance
自己間引き	self-thinning
支持根	prop root
脂質	lipid
子実，穀実	grain
子実体（形成）《病》	fructification
指示物質法	index method
糸状ウイルス	filamentous virus
刺傷［種皮］処理	scarification
自殖，自家受精	self-fertilization, selfing, autogamy
自殖系統	selfed line
自殖性［自家受粉］作物	self-pollinated crop
自殖性植物	autogamous plant
雌穂《トウモロコシ》	ear
雌ずい，めしべ	pistil
雌ずい先熟	protogyny
指数	index
指数関数的成［生］長	exponential growth
指数散布	exponential distribution
システムズアプローチ	systems approach
システム分析	systems analysis
システムモデル	system model
自生地，生育地，生［棲］息地	habitat
自生（の），在来（の）	native, indigenous, spontaneous
雌性不ねん【稔】	female sterility
自生野草	native herb
自然下種	natural reseeding, volunteering
自然乾燥	natural curing
自然乾草	sun-cured hay, field-cured hay
自然群落	natural community
自然交雑	natural crossing
自然交配，まき牛繁殖	free breeding on pasture, pasture breeding
自然雑種	natural hybrid
自然集団	natural population
自然受粉	natural pollination
自然選択，自然とうた【淘汰】	natural selection
自然草地	natural grassland
自然とうた【淘汰】，自然選択	natural selection
自然突然変異	natural mutation
自然保護区	nature reserve

自然保全 [保護]	nature conservation	実容積	actual volume
自然実生, 天然苗	natural seedling	実用品種, 市販品種	commercial variety
飼　槽	trough	自動給餌器	automatic [self] feeder
飼草収量	forage yield		
持続可能な開発	sustainable development	自動給水器	automatic waterer
		自動計測	automatic measurement
子孫, 後代, 次代	progeny, offspring	自動制御	automatic control
次代, 後代, 子孫	progeny, offspring	地ならし機	land leveler
		芝 (地), 芝生	turf, lawn
下草, 下繁草, 下層植生	bottom grass, undergrowth, understory	シバ (型) 草地	Zoysia-type grassland, Zoysia dominated grassland, Zoysia pasture
下　葉	lower leaf		
質, 品質	quality		
湿　害	excess-moisture [humid, wet] injury	柴ハロー	bush harrow
		芝生, 芝 (地)	turf, lawn
失活, 不活性化	inactivation	市販品種, 実用品種	commercial variety
湿気, 湿度	humidity		
湿　原	bog, moor	指標, 指標物質	indicator
実験区, 試験区	experimental plot	指標植物, 植物指標	indicator plant
実験計画, 試験設計	experimental design, design of experiment	耳票付け	ear marking
		指標物質, 指標	indicator
実験ほ【圃】場	experimental field	飼肥料木	tree for green manure and fodder
実効温度, 有効温度	effective temperature		
		師 [篩] 部	phloem
実質利用	actual use	ジベレリン	gibberellin
実態調査	investigation of actual condition, actual condition survey	脂　肪	fat
		死　亡	mortality
		脂肪酸	fatty acid
湿地, 沼地	marsh	脂肪分解酵素, リパーゼ	lipase
湿　地	wetland		
質的形質	qualitative character	脂肪補正乳	fat-corrected milk, FCM
湿　田	ill-drained paddy field		
湿度, 湿気	humidity	死亡率	death [mortality] rate
室内検定	laboratory test	地干し列, 集草列, ウインドロー	windrow
実用 [農業, 経済] 形質	economic [agronomic] character [trait]		
		シミュレーションモデル	simulation model
実用性	practical use		

締固め，圧密	consolidation, compaction	重過リン酸石灰	double superphosphate
霜枯れ草，フォッゲージ	foggage	周期現象，周期性	periodicity
霜柱	frost pillar	周期性，周期現象	periodicity
ジャージー牛	Jersey	周期的発生	periodic occurrence
ジャイロ形テッダレーキ	rotary tedder	重金属	heavy metal
舎飼	housing	重金属汚染	heavy metal pollution
社会性昆虫	social insect	集合，凝集	aggregation
若枝，苗条，茎（部）	shoot	集合，収れん【斂】，収束	convergence
弱小分げつ	weak tiller	自由採食（量）	voluntary intake
若虫	nymph	シュウ酸	oxalic acid
弱毒性，非病原性	avirulence	収支	budget, balance
		抽出，抽出法	extraction
若木	young tree	抽出誤差	sampling error
遮光	shading	抽出物［液］	extract
シャトルベクター	shuttle vector	重しょく【埴】土	heavy clay
しゃへい【遮蔽】，病徴隠ぺい，不顕性の	masking	自由水，遊離水	free water
		集水域	watershed
		集水面積	catchment area
斜面，傾斜	slope	習性	habit
斜面［傾斜］方位	slope aspect	集積，たい【堆】積，蓄積	accumulation
蛇紋岩	serpentinite		
車輪型トラクタ	wheel tractor	自由選択給餌	cafeteria feeding, free choice feeding
種	species, race		
汁液	sap	自由選択法，キャフェテリア法	cafeteria method
汁液接種	sap inoculation		
汁液伝染	sap transmission		
周縁効果	edge [border] effect	終霜	last frost
重回帰	multiple regression	集草列，ウインドロー，地干し列	windrow
臭覚，きゅう【嗅】感覚	olfaction		
収穫機，ハーベスタ	harvester	収束，収れん【斂】，集合	convergence
収穫期	harvest time	従属栄養	heterotrophism, heterotrophy
収穫（する）	harvest		
収穫日	harvesting date		
収穫法	harvest method		

従属栄養生物, 従属栄養体, 有機栄養生物	heterotroph		returns
		重量比	weight ratio
		重力水	gravitational water
従属栄養体, 従属栄養生物, 有機栄養生物	heterotroph	収れん【斂】, 収束, 集合	convergence
		主かん【稈】	main culm
従属種	subordinate species	樹冠	canopy, crown
柔組織	parenchyma	種間競争	interspecific competition
集団, 個体群, 母集団	population	種間交雑	interspecific hybridization
集団遺伝学	population genetics		
集団改良	population improvement	種間雑種	interspecific hybrid
		種間(の)	interspecific
集団[個体群]構造	population structure	種間変異	interspecific variation
		種牛, 雄牛	bull
集団採種	mass seed production	熟期	maturing stage
集団除雄	bulk [mass] emasculation	宿主, ホスト	host
		宿主域[範囲]	host range
集団選抜	mass selection	宿主特異的毒素	host-specific toxin
集中的発芽	clumped germination	宿主範囲[域]	host range
重窒素	heavy nitrogen, ^{15}N	熟成	maturity, maturation, ageing
集中分布	aggregated distribution		
		熟畑	mature field
充填, 詰込み	filling	縮葉	rugose
自由度	degree of freedom	主茎	main stem
収入, 所得	income	主効果, 主作用	main effect
周年栽培	year-round culture	受光係数	light-receiving [intercepting] coefficient
重粘土	heavy soil		
周年(の)	year-round		
周年放牧	yearlong grazing	受光態勢	light-distribution [intercepting] characteristics
重放牧	heavy grazing [stocking]		
		受光量	received light intensity, intercepted light intensity
集約栽培	intensive cultivation		
集約的農業	intensive farming		
集約放牧	intensive grazing		
収率, 収量	yield		
収量, 収率	yield	主根	main [tap] root
重量	weight	主作物	main crop
収量曲線	yield curve	主作用, 主効果	main effect
収量限界	critical yield	主産物	main product
収量構成要素	yield component	種子, 種[たね]	seed
収量漸減の法則	law of diminishing	種子親	seed parent

種子鑑別	seed identification
種子検査	seed inspection [testing]
種子検定 [証明]	seed certification
種子根	seminal root
種子サイズ	seed size
種子散布	seed dispersal
種子重	seed weight
種子収集	seed collection
種子収量	seed yield
種子春化	seed vernalization
種子消毒	seed disinfection
種子証明 [検定]	seed certification
種子処理	seed treatment
種子生産, 採種 (栽培)	seed production
種子生産量, 繁殖率 [力]	fecundity, reproductivity
種子精選	seed clean
種子接種	seed inoculation
種子前処理	seed pretreatment
種子増殖	seed multiplication
種子造粒	seed pelleting
種子貯蔵	seed storage
種実 [種子] ねん【稔】性	seed fertility
種子伝染	seed transmission
種子伝染病	seed-borne disease
種子 [種実] ねん【稔】性	seed fertility
種子発芽	seed germination
種子バンク	seed bank
種子繁殖	seed propagation
種子被覆	seed coating
種子品質	seed quality
種子品質管理	seed quality control
種子粉衣	seed dressing
種子分類基準	seed standards
種子保存	seed preservation
種子量	seed output
種数面積曲線	species-area curve
受　精	fertilization
授　精	insemination
受精能, ねん【稔】性, 繁殖性	fertility
主成分	principal component
種組成《植物》, 植物構成, 草種構成, 草種割合	floristic [botanical] composition
受胎率	conception rate
種多様性	species diversity
手　段	means
種　畜	breeding stock
出芽, ほう【萌】芽	budding, emergence
出芽期	budding time
出　現	appearance, emergence
宿根性, 多年生 (の)	perennial
出　穂	ear emergence, earing, heading
出穂期	heading date [stage, time]
出穂初期	early heading stage
出穂始め	first heading
出生率	birth rate
出　葉	leaf emergence [appearance]
出葉間隔	leaf emergence [appearance] interval
出葉速度	leaf appearance rate
出らい【蕾】	budding
主働遺伝子抵抗性	major gene resistance
受動的吸収	passive absorption
種内競争	intraspecific competition
種内の	intraspecific
授乳, 泌乳, 乳汁分泌	lactation

授乳期, 泌乳期間	lactation period	純生産	net production
授乳牛, 子付（繁殖）雌牛	suckler cow	純タンパク【蛋白】質	true protein
種の豊富さ	species richness	じゅん【馴】致	accustoming, acclimation, preconditioning
種 皮	seed coat		
樹 皮	bark	じゅん【馴】致放牧	warm-up grazing
種皮［刺傷］処理	scarification	純度, 精選度	purity
樹皮たい【堆】肥, バークたい【堆】肥	bark compost	純同化率	net assimilation rate, NAR
		順 応	accomodation
種 苗	seeds and seedlings	子 葉	cotyledon
種苗登録	registration of plant varieties for the Seeds and Seedlings Law	飼養, 摂食	feeding
		消化, 温浸	digestion
		浄 化	purification
		乗が【駕】	mounting
受粉, 送粉, 花粉ばい【媒】介	pollination	障害物	barrier
		消化管	alimentary canal, digestive tract
主脈《植物》	rib	消化器	digestive organ
寿命, 長寿	longevity	硝化菌 ［硝酸化成菌］	nitrifying bacteria, nitrifier
主要種	key species		
受容力	capacity	硝化作用	nitrification
狩 猟	hunting	消化試験	digestion trial
樹林地, 林地, 森林	woodland, forestland	浄化処理	purification treatment
		消化性, 消化率	digestibility
春化, バーナリゼーション	vernalization	消化速度	digestion rate
		小花［果］柄, 小枝こう【梗】	pedicel
じゅん【馴】化, 順化	acclimation, acclimatization, domestication		
		消化率, 消化性	digestibility
		飼養管理	feeding management
順化, じゅん【馴】化	acclimation, acclimatization, domestication	小根, 幼根	radicle
		常在度, 恒存度	constance
		上作, 豊作	abundant [good] harvest
循環選抜	recurrent selection		
純寄生者［菌］, 絶対寄生菌	obligate parasite	硝 酸	nitric acid
		蒸 散	transpiration
純 系	pure line	硝酸アンモニウム	ammonium nitrate
純系選抜	pure line selection		
純光合成, 見かけの光合成	net [apparent] photosynthesis	硝酸塩	nitrate
		硝酸（塩）中毒	nitrate poisoning

し [37]

硝酸還元	nitrate reduction
硝酸還元菌	denitrifying [nitrate reducing] bacteria
硝酸還元酵素	nitrate reductase
硝酸菌	nitrite oxidizing bacteria
硝酸植物	nitrate plant
蒸散速度	transpiration rate
硝酸態窒素	nitrate nitrogen
飼養試験	feeding experiment
小枝こう【梗】, 小花［果］柄	pedicel
消失速度, 消失率	disappearance rate
壌質砂土	loamy sand
消失率, 消失速度	disappearance rate
照 射	irradiation
照射, 放射, 放射線	radiation
蒸 煮	steaming
子葉しょう【鞘】, しょう【鞘】葉	coleoptile
小 穂	spikelet
脂溶性ビタミン	fat soluble vitamin
焼成りん肥	calcined phosphate
消石灰	slaked lime
上層採食, トップグレージング	top grazing
状態診断	condition diagnosis
照度, 光強度	light intensity, illuminance
壌 土	loam
少糖, オリゴ糖	oligosaccharide
消 毒	disinfection
小農具	small implement
条 播	row [stripe] seeding [sowing]
上はい【胚】軸	epicotyl
条播機, ドリル	drill
蒸 発	evaporation
蒸発散	evapotranspiration
小板, 盤状体, はい【胚】盤	scutellum
上繁草	top grass
消費者	consumer
飼養標準	feeding standard
小 苞	bractlet
小胞体	endoplasmic reticulum [pl. endoplasmic reticula], ER
静脈, 脈	vein
しょう【鞘】葉, 子葉しょう【鞘】	coleoptile
小 葉	leaflet
照葉樹林帯	laurel forest zone
蒸留法	distillation method
省力化, 省力（の）	labor saving
省力管理	labor saving management
省力（の）, 省力化	labor saving
常緑の	evergreen
奨励品種	recommended variety
初期生育	early growth
初期世代	early generation
除 去	removal
食塩, 塩化ナトリウム	sodium chloride, salt
食 害	defoliated [feeding] damage
しょく【埴】壌土	clayey loam
食植性昆虫	phytophagous insect
食 性	feeding habit
植 生	vegetation
植生構造	vegetation structure
植生交替	vegetational

[38] し

	replacement
植生図	vegetation map
植生推移	vegetational transition
植生遷移	plant succession
植生動態	vegetation dynamics
植生パターン	vegetation pattern
植生連続	vegetational continuum
しょく【埴】土, 粘土	clay
食道	esophagus
植被	plant cover [coverage]
植物	plant
植物育成箱, グロースキャビネット	growth cabinet
植物栄養	plant nutrition
植物解剖学	plant anatomy
植物学	botany
植物群落	plant community
植物形態学	plant morphology
植物構成, 草種構成, 草種割合, 種組成《植物》	floristic [botanical] composition
植物個体群統計学	plant demography
植物色素	plant pigment
植物指標, 指標植物	indicator plant
植物社会学	plant sociology
植物生態学	plant ecology
植物生理学	plant physiology
植物相, フローラ	flora
植物組織	plant tissue
植物組成	plant composition
植物帯	floral zone
植物体	plant body
植物体の形状	plant shape
植物多様性	floral diversity
植物毒性, 殺草性	phytotoxicity
植物の機能型［タイプ］	plant functional type
植物の交替［入替わり］	plant turnover
植物のコロナイゼーション［すみつき］	plant colonization
植物品種保護	plant variety protection
植物分布様式	plant distribution pattern
植物ホルモン	phytohormone, plant hormone
植物養分	plant nutrient
食物環	food cycle
食物網	food web
食物連鎖	food chain
食欲	appetite
植林, 人工造林	afforestation
植林地, 造林地	plantation, afforested land
処女［単為］生殖	parthenogenesis
除草	weeding
除草機	weeder
除草剤	herbicide
除草剤耐性	herbicide tolerance [resistance]
ショ糖, スクロース	sucrose, saccharose
所得, 収入	income
初乳	colostrum
暑熱	hot temperature, heat
除雄《植物》, 去勢《動物》, 性腺摘除《動物》	castration, emasculation
処理, 治療	treatment
白子, アルビノ, 白色種	albino

シリカ, 無水ケイ酸	silica		[impact]
人為的突然変異, 人工突然変異		artificial mutation	
自立経営	viable farm	進化	evolution
飼料, 餌	feed (stuff), diet, fodder	真菌類, かび類, 菌類	fungus [*pl.* fungi]
試料, 標本, 検体	sample	真空乾燥機	vacuum dryer
飼料運搬車	feed carrier	真空サイロ	vacuum silo
飼料学	feed science	真空ポンプ	vacuum pump
飼料価値	feeding value	深耕	deep plowing
飼料鑑定	judgement of feed	人口学, 個体群統計学, デモグラフィ	demography
飼料構造	structure of feed utilization		
飼料効率	feed efficiency	人工乾燥	artificial drying
試料採取, 標本抽出	sampling	人工気象室	climatic chamber, air conditioning room
飼料作物	forage crop	人工気象室《植物用》, ファイトトロン	phytotron
飼料臭	feed flavor		
飼料成形機	feed pelleting machine, pelleting machine		
		人工群落	artificial community, cultivated vegetation
飼料生産	forage production		
飼料精選機	feed cleaner		
飼料成分組成	composition of feed, feed composition	人工交雑, 人工交配	artificial crossing
飼料摂取（量）	feed intake	人工交配, 人工交雑	artificial crossing
飼料選択, 餌選択	diet selection		
		人工種子	artificial seed
飼料草	forage herb	人工授精	artificial insemination, AI
飼料組成	ingredient composition		
		人工授粉	artificial pollination
飼料添加物	feed additive	人工消化率	*in vitro* [artificial] digestibility
飼料配合機	feed mixer		
飼料畑	forage crop field	人工照明	artificial lighting
飼料品質	feed quality	人工接種	artificial inoculation
飼料分析	feed analysis	進行遷移	progressive succession
飼料補給, 補助飼料給与	supplemental [supplementary] feeding		
		人工草地	sown [artificial] grassland
飼料木	fodder tree [shrub]	人工造林, 植林	afforestation
飼料要求量	feed requirement	進攻の《生態》	aggressive
人為圧	artificial pressure		

人工突然変異, 人為的突然変異	artificial mutation	侵入者	invader
		侵入植物	invading [alien] plant
		心拍数	heart rate
人工乳	synthetic milk	真比重, 比重	specific gravity
深耕プラウ	deep plow	新芽（をくう）	browsing
人工哺乳	artificial suckling	針葉樹	conifer
深根性作物	deep-rooted crop	針葉樹林	coniferous forest
新墾プラウ	breaker [breaking] plow	信頼区間	confidence interval
		森林, 樹林地, 林地	woodland, forestland
浸種	seed presoaking		
浸出, 洗脱, リーチング	leaching	森林土壌	forest soil
		森林伐採	forest clearance, deforestation
浸食	erosion《衝撃などによる》		
		親和性	affinity, compatibility
侵食	erosion《水・風などによる》		
		す	
浸漬	soaking		
新鮮重《植物》	fresh weight	推移行列	transition matrix
新鮮物	fresh matter	スィープレーキ	sweep [buck] rake
新鮮ふん【糞】, 生ふん【糞】	fresh feces	水牛, バッファロ	buffalo
深層[下層]施肥	deep placement, deep application of fertilizer	水銀	mercury, Hg
		穂型	panicle [ear, head, inflorescence] type
診断	diagnosis	水耕	hydroponics, solution [water] culture
伸長, 延長	elongation, extension		
伸長帯	growing region		
心土, 下層土	subsoil	水酸化ナトリウム	sodium hydroxide
浸透	penetration		
浸透圧	osmotic pressure	穂軸	cob, rachis
浸透現象	osmosis	水質	water quality
浸透性, 透過性, 透過率	permeability	水質汚染 [汚濁]	water pollution
		水質汚濁 [汚染]	water pollution
心土 [土層] 改良	subsoil improvement	衰弱, 減退, 低下	decline
心土耕	subsoil plowing	水準	level
心土破砕	subsoiling	穂状花序, 花穂, 穂	spike
心土破砕機	pan-breaker		
侵入・定着, すみつき	colonization	水食	water erosion
		水浸状の	water-soaked
侵入・定着能力, すみつき能力	colonizing ability	水生雑草	aquatic [hydrophytic] weed

水生植物	hydrophyte	水溶性ビタミン	water soluble vitamin
水生バイオマス	aquatic biomass	水溶性リン酸	water soluble phosphate
水　素	hydrogen, H		
水素イオン濃度	hydrogen ion concentration	水路，チャンネル	channel
衰退，荒廃，劣化	deterioration, degradation	水和剤	wettable powder
		すき	plow
衰退［荒廃］草地	degrading grassland, waste grassland, deteriolated grassland	スキャン幅	scan width
		すくみ，発育阻止，わい【矮】化	stunt
穂　長	ear length	スクリーニング	screening
垂直茎，直立茎	orthotropic stem ［shoot］, erect stem ［shoot］	スクロース，ショ糖	sucrose, saccharose
		ススキ型草地	Miscanthus-type ［silvergrass-type］ grassland
垂直分布	vertical distribution		
推　定	estimation		
推定値	estimate, estimated value	スタック，野積み	stack
水　田	paddy field	スタックサイロ	stack silo
水田裏作	winter cropping on drained paddy field	スタンチョン	stanchion
		スタンド	stand
水田転換畑	converted paddy field	スタンプ，根株《樹の》	stump
水田土壌	paddy soil		
水田利用再編	reorganization on paddy field use	スタンプカッタ	stump cutter
		ステップ《中央アジアの内陸草地》	steppe
出　納	balance		
水　分	moisture		
水分含量，含水量	moisture [water] content	ステロイド	steroid
		ストール，牛床	stall
水分当量	moisture equivalent	ストリップ放牧	strip grazing
水分平衡	water balance	ストレス	stress
水分飽和度	relative turgidity	ストロマ《クロロプラストの》	stroma [*pl.* stromata]
水平分布	horizontal distribution		
水溶性炭水化物 ［糖類］	water soluble carbohydrate, WSC	スプリンクラーかんがい【灌漑】	sprinkler irrigation
水溶性タンパク 【蛋白】質	water soluble protein		
		スペーサ	spacer
水（溶性）抽出物	water extract	滑り率	slippage
		すみつき，侵入・定着	colonization
水溶性糖類 ［炭水化物］	water soluble carbohydrate, WSC		

すみつき能力, 侵入・定着能力	colonizing ability
すみわけ	habitat segregation, separation
スラグ, 鉱さい	slag
スラッジ, 汚泥	sludge
スラリー, 泥状物	slurry
スラリーインジェ[ゼ]クタ	slurry injector
スラリー散布機	slurry spreader
スラリーポンプ	slurry pump

せ

性	sex
生育, 成[生]長	growth and development, growth
生育形, 生育型, 成[生]長型	growth form, growth type
生育型, 成[生]長型, 生育形	growth form, growth type
生育過程	growing process
生育期間, 成[生]長期間, 生育時期	growing season
生育時期, 成[生]長期間, 生育期間	growing season
生育習性, 成[生]長習慣, 生育特性	growth habit
生育促進, 成[生]長促進	growth stimulation, growth promotion
生育地, 自生地, 生[棲]息地	habitat
生育特性, 生育習慣, 成[生]長習慣	growth habit
生育日数	growing period
生育量	amount of growth
生化学, 生物化学	biochemistry
生活型[形]	life form
生活形スペクトル, 生活型組成	biological spectrum
生活型組成, 生活形スペクトル	biological spectrum
生活環	life cycle
生活形[型]	life form
生活史	life history
正規曲線	normal curve
正規分布	normal distribution
正規偏差	normal deviation
正規母集団	normal population
正逆交雑	reciprocal crossing
制御実験[試験]	controlled experiment
静菌作用	fungistasis
成形乾草	compressed [processed] hay
制限アミノ酸	limiting amino acid
制限給餌	controlled [restricted] feeding
制限酵素断片長多型	RFLP [restriction fragment length polymorphism]
制限的資源	limiting resource
生合成	biosynthesis
成(個)体, 成虫	adult, imago
青酸	hydrocyanic acid
生産, 生成	production
生産エネルギー	production energy
生産技術	production technique
生産計画	production plan
生産効率	production efficiency
生産者, 産生株	producer
生産性, 生産力	productivity, performance

生産費	productive expenses, production cost	生存株	surviving plant [ramet]
生産量	production	生存競争	struggle for existence
生産力, 生産性	productivity, performance	生存曲線	survivorship [survival] curve
生産力検定	performance test	生存年限	persistency
静止期, 誘導期	lag phase	生存率	survival ratio
性周期, 発情周期	estrous cycle	生体外の, インビトロ, 試験管内の	in vitro
成 熟	maturation	生態学	ecology
成熟期	ripening [maturity] stage	生態型, エコタイプ	ecotype, biotype
成熟個体《植物》	mature plant	生態型選抜	ecotype selection
成熟はい【胚】	mature embryo	生態系	ecosystem
成熟[減数, 還元]分裂	meiosis	生体恒常性, ホメオスタシス	homeostasis
生殖《動物》, 繁殖《動物》	reproduction, breeding	生態種	ecospecies
生殖質	germplasm	生体重《動物》	live weight
生殖成[生]長期, 繁殖期	reproductive stage	生態遷移	ecological succession
生成, 生産	production	生態的最適性	ecological optimum
生石灰	quick lime	生態的特性	ecological attribute, ecological trait
性腺摘除《動物》, 除雄《植物》, 去勢《動物》	castration, emasculation	生態的プロセス	ecological process
		生体内の, インビボ	in vivo
精選度, 純度	purity	生態ピラミッド	ecological pyramid
生 草	green forage	生態分布	ecological distribution
生草[青刈]給与, 青刈り	soiling, fresh forage feeding, zero grazing	成帯[帯状]分布, ゾーネーション	zonation
生草[青刈]収量	fresh [green] yield	整 地	land [soil] preparation, leveling of ground
生[棲]息地, 自生地, 生育地	habitat	静 置	standing
		成虫, 成体, 成個体	adult, imago
生[棲]息地分断化	habitat fragmentation	成[生]長, 生育	growth and development, growth
生[棲]息密度, 個体数[群]密度	population density		
生 存	survival, survivorship	成[生]長解析	growth analysis

成[生]長型, 生育型, 生育形	growth form, growth type
成[生]長期 成[生]長期間, 生育時期, 生育期間	growth period growing season
成[生]長曲線	growth curve, increment curve of growth
成[生]長習慣, 生育習性, 生育特性	growth habit
成[生]長相, 増殖相	growth phase
成[生]長阻害物質	growth inhibitor
成[生]長促進, 生育促進	growth stimulation, growth promotion
成[生]長速度	growth rate
成[生]長段階	growth stage
成[生]長調節物質	growth regulator
成[生]長点, 分裂組織	meristem
成[生]長点	growing point
成[生]長ホルモン	growth hormone
成[生]長様式	growth pattern
成[生]長抑制剤[物質]	growth retardant
西南暖地	south-western warm region
生 乳	raw milk
性能[能力]検査	performance test
生物価	biological value, BV
生物化学, 生化学	biochemistry
生物化学的酸素要求量	biochemical oxygen demand, BOD
生物(学)的防除[除去]	biological control
生物環境修復, バイオリメディエーション	bioremediation
生物気候	bioclimate
生物季節学, フェノロジー	phenology
生物共同体	biotic community
生物計	biometer
生物経済学	bioeconomics
生物(学的)検定, バイオアッセイ	bioassay
生物圏, バイオスフェア	biosphere
生物指標	biotic indicator
生物生態学	bioecology
生物相	biota
生物相互作用	coaction
生物測定学	biometrics
生物(体)量, バイオマス	biomass
生物多様性	biodiversity
生物要因	biotic factor
成 分	ingredient
成雌牛	cow
生理化学的	physiochemical
生理学的	physiological
生理障害	physiological disorder
生理食塩水, 塩(性)水	saline water
生理生態学的	physioecological
生理的塩基性肥料	physiologically [potentially] basic fertilizer
生理的統合	physiological integration
生理的特性	physiological characteristics
生理的分化, 寄生性分化	physiological specialization

せ [45]

生理病	physiological disease	カルシウム沈着	
ゼオライト, ふつ【沸】石	zeolite	石灰植物	calciphilous plant, calcicole
赤黄色土	red-yellow soil	石灰窒素	calcium cyanamide
赤外線ガス分析計	infrared gas analyzer	節間伸長	internode elongation
積算温度	accumulated [cumulative] temperarure	節間（部）	internode
		積極的吸収, 能動的吸収	active absorption
積算気温	accumulated [cumulative] air temperature	接合, 交配, 交尾	mating
		接合［交配］型	mating type
積算降水量	accumulated precipitation	節根	nodal root
		接種	inoculation
積算日射量	integrated solar radiation	摂取，経口摂取	ingestion
		接種吸汁	inoculation feeding
積算優占度	summed dominance ratio, SDR	接種源	inoculum [*pl.* inocula]
		摂取［採食］行動	grazing [ingestive] behavior
積雪地帯	snow zone （area, region）		
		摂取量	intake
施工管理	execution management	摂食，飼養	feeding
		絶食, 空腹時	fasting
施工基準	execution standard	節足動物相	arthropod fauna
施工法	execution method	絶対寄生菌, 純寄生者［菌］	obligate parasite
世代	generation		
世代間隔	generation interval	切断	cutting
世代交代	alternation of generations, heterogenesis	切断機構	cutting mechanism
		切断長	chop length
		切断法	cutting method
世代促進	accelerated generation advancement	切断葉	detached leaf
		節密度	node density
節	node	絶滅	extinction
節位	node order	絶滅危ぐ【惧】（植物）種	endangered (plant) species
切えい【穎】法	clipping method		
石灰	lime	施肥	fertilizer application, fertilizing, fertilization
雪害	snow damage		
石灰岩	limestone		
石灰散布機	lime sower	施肥機	fertilizer distributor
石灰質土壌	calcareous soil	施肥時期	time of fertilizer application
石灰質肥料［資材］	liming material, calcium fertilizer		
		施肥播種機	fertilizer drill
石灰集積作用,	calcification	施肥標準	fertilizing standard

施肥法	method of fertilizer application		grazing
		全固形物	total solid body
施肥量	application rate of ferlilizer, fertilizer rate	せん【剪】根	root pruning
		潜在［潜伏］感染	latent infection
施用	application	潜在地力	potential soil fertility [productivity]
競合い，競合［争］	competition		
		前作	preceding cropping
セリン	serine	前作物	preceding crop
セルフフィーダ	self-feeder	選種	seed grading
セルラーゼ	cellulase	扇状地，沖積扇状地	alluvial fan
セルロース	cellulose		
遷移，サクセッション	succession	線状頻度法	string method
		染色質	chromatin
線維，繊維	fiber	染色体，クロモソーム	chromosome
繊維，線維	fiber		
前胃《反すう【芻】動物》	forestomach	染色体数	chromosome number
		染色体地図	chromosome map
遷移後期種	late-successional species	染色体突然変異	chromosomal mutation
繊維根，ひげ根	fibrous root	染色分体	chromatid
繊維質飼料	fibrous feed	前処理［処置］	pretreatment
遷移初期種	early-successional species	全身感染	systemic infection
		全身病徴	systemic symptom
遷移段階	stage of succession	選択吸収	selective absorption
遷移度	degree of succession	選択採食	selective grazing [herbivory], dietary selection
繊維分解酵素	fiber degrading [breakdown] enzyme		
		選択順位	preference ranking
遷移平衡	successional equilibrium	選択性除草剤	selective herbicide
		選択培地	selective medium
前期，分裂前期	prophase	洗脱，浸出，リーチング	leaching
専業農家	full-time household		
先駆［前駆］物質［体］	precursor	全炭素	total carbon
		せん【剪】断抵抗	shear resistance
前駆［先駆］物質［体］	precursor		
		全短波放射，全天日射	global [total solar, total short wave] radiation
前駆牧草	precursor forage plant		
線形計画	linear programing		
線形模型	linear model	全窒素	total nitrogen
浅耕	shallow tillage	線虫	nematode, eelworm
先行後追放牧	leader-follower	線虫病	nematode disease

全天日射, 全短波放射	global [total solar, total short wave] radiation	相加作用	
		草冠, 林冠, キャノピ	canopy
せん【尖】度	kurtosis	相　観	physiognomy
全　糖	total sugar	相　関	correlation
全　乳	whole milk	相関係数	correlation coefficient, coefficient of correlation
潜　熱	latent heat		
選　抜	selection		
選抜基準	selection criterion		
選抜効果	selection effectiveness	草冠構造	canopy structure
選抜法	selection method	相関分析	correlation analysis
せん【尖】部, 頂部, 頂端	apex	早期検定	early diagnosis
		早期栽培《水稲》	early season culture
潜伏［潜在］感染	latent infection	早期播種, 早播き	early seeding [sowing]
潜伏期間	incubation [latent] period	早期離乳	early weaning
		草原, 草地	grassland
全面散布	overall application	草原植物	glassland plant
全面処理《除草剤の》	overall treatment	草　高	plant [sward] height
		走光性	phototaxis
せん【剪】葉, 落葉, 摘葉処理	defoliation	総合防除	integrated control
		相互関係	interaction, interrelationship
千粒重	thousand kernel [seed] weight	相互作用, 交互作用（項）	interaction
線量, 投与量, 服用量	dose	相互しゃへい【遮蔽】	mutual shading
		操作, ハンドリング	operation, handling
そ			
		掃除刈り	trimming
層, 階層	stratum	草　種	herb species
総当たり交配［交雑］, 二面交雑［交配］	diallel crosses	増　収	yield increase
		草種競合	competition of grass, interspecific competition of grass
総エネルギー	gross energy		
桑　園	mulberry field [plantation]	草種構成, 植物構成,	floristic [botanical] composition
霜　害	frost injury	草種割合, 種組成《植物》	
相加作用, 相加的効果	additive effect		
増加速度（率）	rate of increase		
相加的効果,	additive effect		

草種割合, 植物構成, 草種構成, 種組成《植物》	floristic [botanical] composition		gain
		相対湿度	relative humidity
		相対重量	relative weight
		相対照度	relative light intensity
総状花序	raceme	相対成[生]長	allometry, relative growth
そう【叢】状型	bunch [tufted] type		
そう【叢】生型, 株型		相対成[生]長率	relative growth rate, RGR
相乗効果	multiplier effect	草地	grassland, sward
双子葉植物, 双子葉類	dicotyledon	草地, 草原	grassland
		草地開発	grassland development
双子葉類, 双子葉植物	dicotyledon	草地改良[改善]	grassland [pasture] improvement
増殖《微生物の》	growth		
増殖	multiplication	草地学	grassland science
増殖《植物》	propagation	草地型	grassland type
増殖相, 成[生]長相	growth phase	草地簡易改良	simple improvement of grassland
草食動物	herbivore, herbivorous animal	草地管理	grassland management
そう【叢】生, ロゼット	rosette	草地群落	grassland community
		草地景観	grassland landscape
草勢	plant vigor	草地更新	pasture renovation
造成	establishment	草地雑草	pasture [grassland] weed
草生改良	improvement of grassland	草地植生	grassland vegetation
そう【叢】生型, そう【叢】状型, 株型	bunch [tufted] type	草地診断	grassland diagnosis
		草地生産	grassland [pasture] production
蔵精器	antheridium [pl. antheridia]	草地生態学	grassland ecology
		草地生態系	grassland ecosystem
造成工法	establishment method, method of grassland reclamation	草地造成	pasture establishment
		草地畜産	grassland farming [husbandry]
		草地土壌	pasture [grassland] soil
草生栽培	sod-culture		
総生産(量)	gross production	草地農業	grassland agriculture [farming]
造成地	established [reclaimed] land	草地保全	grassland conservation
造成法	method of (grassland) establishment		
		草地酪農	grassland dairying
増体, 体重増加	body [live] weight	草地利用	grassland utilization

そ [49]

相同染色体	homologous chromosome	側 根	lateral root
双はい【胚】植物	twin plant	側 枝	lateral shoot
		測 定	measurement
早晩性	earliness	測定値	measured value, data
送風機, ブロワ	blower	側方施肥	side manuring
送粉, 受粉, 花粉ばい【媒】介	pollination	そ【蔬】菜, 野菜	vegetable
		組 織	tissue
		組織化学的	histochemical
送粉者, ポリネータ, 花粉ばい【媒】介者	pollinator	組織学	histology
		組織学的	histological
		組織培養	tissue culture
		粗脂肪	crude fat
層別化, 階層構造	stratification	そしゃく【咀嚼】	mastication, chewing
		そしゃく【咀嚼】時間	chewing time
層別刈取法	stratified clip method		
草 本	herb, herbaceous plant	疎 植	sparse planting
相利共生, 共生	symbiosis, mutualism	粗飼料	bulky feed, roughage
草 量	herbage mass	組 成	constituent, composition
草量計	grass [pasture] meter, electronic capacitance probe	塑性限界	plastic limit
		粗繊維	crude fiber
造林地, 植林地	tree plantation, afforested land	粗大有機物	crude undecomposed organic matter
装輪トラック, 輪だち【轍】	wheel track [rut]	粗タンパク【蛋白】質	crude protein
ソース・シンク関係	source-sink relation	速効性肥料	quick acting manure, readily available fertilizer
ゾーネーション, 帯状[成帯]分布	zonation	粗放管理	extensive [rough] management
ゾーン電気泳動, 分域電気泳動	zone electrophoresis	粗放な	extensive
		粗放牧	rough [lax] grazing
阻 害	inhibition	ソマクローナル[体細胞]変異	somaclonal variation
阻害剤, インヒビター	inhibitor	損益計算	profit and loss method
粗灰分	crude ash	損 失	loss
属	genus	損 傷	damage
側 芽	lateral bud		
属間交雑	intergeneric crossing		
属間(交)雑種	intergeneric [genus] hybrid		

た

ターンオーバー, 回転	turnover
ダイアレル分析	diallel analysis
第一・二胃, 反すう【芻】胃	reticulorumen
第一胃, 反すう【芻】胃, ルーメン	rumen
第一胃［ルーメン］内細菌	rumen bacteria
第一胃［ルーメン］内発酵	rumen fermentation
第一胃内溶液	rumen fluid
第一胃内原生動物	rumen protozoa
第一種兼業農家	farm-household mainly engaged in farming
耐陰性	shade tolerance
耐塩性	salt resistance [tolerance]
ダイオキシン	dioxin
体温	body temperature
退化, 変性, 逆行変性	degeneration
台芽, ひこばえ	sucker
耐寒性, 耐冷性	cold hardiness, cold resistance [tolerance]
耐乾［干, 旱］性	drought resistance [tolerance]
大気汚染	air pollution
大気候	macroclimate
待期放牧	deferred grazing
待期牧区	deferred pasture
大規模な, 広範な	extensive
大気モデル	atmospheric model
耐久性	durability
たいきゅう【堆厩】肥	compost and barnyard manure
待期輪換放牧	deferred rotational grazing
体系	system
退行	regression
退行遷移	retrogressive succession
体細胞雑種	somatic hybrid
体細胞はい【胚】形成	somatic embryogenesis
体細胞［ソマクローナル］変異	somaclonal variation
第三胃	omasum
第三紀層	Tertiary
耐酸性	acid tolerance
胎仔, 胎児	fetus
胎児, はい【胚】	embryo
胎児, 胎仔	fetus
耐湿性	wet endurance
代謝	metabolism
代謝エネルギー	metabolizable energy, ME
代謝生成物, 代謝（産）物	metabolite
代謝（産）物, 代謝生成物	metabolite
代謝体重	metabolic live weight
体重増加, 増体	body [live] weight gain
対照区	control plot
代償性発育	compensatory growth
対照品種, 比較品種	check variety
耐暑性, 耐熱性	heat resistance [tolerance]
耐水性	water resisting property

耐水性団粒	water-stable aggregate	コンポスト化, 発酵乾燥	
対数変換	logarithmic transformation	たい【堆】肥散布機	manure spreader
対数目盛	logarithmic scale		
大豆かす【粕】	soybean meal	耐肥性	adaptability to heavy dressing of fertilizer
耐性, 抵抗性, 抵抗力	resistance, tolerance	たい【堆】肥積込機	manure loader
たい【堆】積, 集積, 蓄積	accumulation	耐病性, 病害抵抗性	disease resistance [tolerance]
たい【堆】積岩	sedimentary rock	耐病性検定圃	disease garden
耐雪性	snow endurance	台風	typhoon
耐霜性	frost hardiness, frost resistance [tolerance]	大便, ふん【糞】, ふん【糞】便	feces
代替農業	alternative agriculture [farming]	退牧	end of grazing, housing from pasture, closing the grazing
代替比 [率]	substitution rate		
台地 [畑] 土壌	upland soil	滞牧日数, 放牧期間	grazing period
耐虫性	insect resistance [tolerance]	太陽エネルギー	solar energy
大腸菌	coliform bacilli	代用乳	milk replacer
多遺伝子, ポリジーン, 微動遺伝子	polygene	耐用年数	durable years, useful life
		第四胃	abomasum
耐踏圧性, 踏圧耐性	trampling tolerance [resistance]	第四紀層	Quarternary
		対立遺伝子, アレル, アレレ	allele
耐冬性	winter hardiness		
耐凍性	freezing hardiness		
耐倒伏性, 倒伏抵抗性	lodging resistance	対流	convection
		滞留	retention
体内分布	internal distribution	滞留時間	retention time
第二胃, 細網《組織》	reticulum	耐冷性, 耐寒性	cold hardiness, cold resistance [tolerance]
第二種兼業農家	farm-household mainly engaged in other jobs	多雨地帯	humid region
		だ【唾】液	saliva
耐熱性, 耐暑性	heat resistance [tolerance]	高うね【畦】	high ridge
		高刈り	high level cutting
大農具	large implement	他家受精	cross fertilization
たい【堆】肥	farmyard manure	他家受粉	allogamy, cross pollination
たい【堆】肥化,	composting		

[52] た

他家受粉作物, 他殖性作物	cross pollinating crop
他家生殖, 他殖	allogamy
他感作用, アレロパシ	alleropathy
濁度	turbidity
たく【托】葉	stipule
多型 (現象)	polymorphism
多系交雑	multiple cross [crossing]
多交配	polycross
多収	high yield
多収 (穫) 栽培	high yielding culture
多重［重複］感染	multiple infection
多収性	high yielding ability
多汁性	juiciness
他殖, 他家生殖	allogamy
他殖性, 他配	outcrossing, outbreeding, allogamy
他殖性作物, 他家受粉作物	cross pollinating crop
他殖性植物	allogamous plant
他殖性 (の)	allogamous
多数性, 倍数体	polyploid
立枯れ	damping-off, standing dead (material)
脱イオン水, 脱塩水	deionized water
脱塩作用	desaltification
脱塩水, 脱イオン水	deionized water
脱穀	threshing
脱脂粉乳	dried skim milk
脱色	decolorization
脱水 (症)	dehydration
脱水素酵素, デヒドロゲナーゼ	dehydrogenase
脱炭酸酵素	decarboxylase
脱窒菌	denitrifying bacteria
脱窒作用	denitrification
脱リグニン	delignification
脱粒	shattering
脱粒性	shattering habit
たてば, 休息場	resting place
多糖 (体)	polysaccharide
棚田	terraced paddy field, rice terrace
ダニ熱	tick fever
種［たね］, 種子	seed
たねまき, 播種	seeding, sowing
多年生 (の), 宿根性	perennial
他配, 他殖性	outcrossing, outbreeding, allogamy
多肥栽培	heavy manuring culture
多変量解析	multivariate analysis
多面的土地利用	multiple landuse
多毛作	multiple cropping
多葉性	leafiness
多量元素［要素］	macroelement, macronutrient, major element
多量要素［元素］	macronutrient, macroelement, major element
多連プラウ	gang plow
タワーサイロ, 塔サイロ	tower silo
単為［処女］生殖	parthenogenesis
単一経営	specialized farming
単胃動物	monogastric animal
単位面積当り放牧頭数	stocking rate
暖温帯	warm temperate zone
弾丸暗きょ【渠】	mole drain

短期輪換放牧	short time rotation [short duration, rapid rotation] grazing	単相体, 半数体, 一倍体	haploid
		炭素源	carbon source
		炭素シンク	carbon sink
短茎イネ科草, 短草型草種	short grass	炭素窒素比, C-N比	C-N ratio, carbon-nitrogen ratio
単交配	single cross		
単細胞タンパク【蛋白】質	single cell protein, SCP	炭素同位体分別	carbon isotope discrimination
探　索	searching, exploitation	担体, キャリア, 保毒[菌]者	carrier
単　作	single cropping	暖　地	warm region
探索時間	search time	暖地型(の), 熱帯の	tropical
炭酸ガス, 二酸化炭素	carbon dioxide [CO_2]		
		単糖(類)	monosaccharide
		断頭, 頂部除去	decapitation
炭酸カルシウム	calcium carbonate	タンニン	tannin
炭酸固定	carbon dioxide [CO_2] fixation	タンニン酸	tannic acid
		短年生草	short-lived grass
炭酸同化	carbon dioxide [CO_2] assimilation	短年草	temporary grass
		短年放牧地	ley pasture
単式遺伝子型	simplex genotype	短年用牧草	pasture plant for short-term use
担子菌(類)	basidiomycetes		
短日植物	short-day plant		
短日処理	short-day treatment	短年輪作草地, 輪作草地	ley
担子柄	basidium [*pl.* basidia]		
担子胞子	basidiospore	単　播	single-crop seeding [sowing]
胆　汁	bile		
単子葉植物	monocotyledon	タンパク【蛋白】価	gross protein value, GPV
単少糖	mono- and oligo-saccharide		
		タンパク【蛋白】合成	protein production
湛水, 洪水	flooding, waterlogging, submergence		
		タンパク【蛋白】合成能	protein production rate
炭水化物	carbohydrate		
炭水化物代謝	carbohydrate metabolism	タンパク【蛋白】効率	protein efficiency ratio, PER
炭　素	carbon, C	タンパク【蛋白】質	protein
短草型	short grass type		
短草型草種, 短茎イネ科草	short grass	タンパク【蛋白】質飼料	protein supplement
		タンパク【蛋白】質分解	proteolysis
短草型放牧草地	short grass pasture		
単(純)相関	simple correlation		

タンパク【蛋白】質分解酵素, プロテアーゼ	protease	地下ほふく【匍匐】枝	underground creeper
タンパク【蛋白】態窒素	protein nitrogen	置換, 入替え	replacement
		置換, 排出	displacement
単播草地	pure stand	置換性塩基, 交換性塩基	exchangeable base
段畑	terrace field	地球温暖化	global warming
ダンプトレーラ	dump trailer	畜産, 畜産経営	animal husbandry, livestock farming
ダンプレーキ	dump rake		
単葉	simple leaf	畜産学	animal science, zootechnology, zootechnical science
単用プラウ	common plow		
単離	isolation		
団粒化	aggregation		
団粒構造	crumbled [aggregate] structure	畜産業	animal [livestock] industry
団粒分析	analysis of water stable aggregate	畜産経営, 畜産	animal husbandry, livestock farming
		畜産物	animal product
ち		畜舎	barn, stable
		蓄積, 集積, たい【堆】積	accumulation
チアミン, ビタミン B_1	thiamin, vitamin B_1		
		畜肥, きゅう【厩】肥, 肥[こえ]	stable [barnyard] manure, manure
地域性, 局所性	locality		
チェーンコンベア	chain conveyer	地形	topography
		地形要因	physiographic factor
チェーンハロー	chain harrow	遅効性	slow [delayed] release fertilizer
地温	soil temperature	[緩効性]肥料	
地下かんがい【灌漑】	sub-irrigation	致死温度	thermal death point
		致死濃度	lethal concentration
地下器官型	radicoid form	地上植物	phanerophytes
地下茎, 根茎	subterranean stem, rhizome	地上部	aboveground [top, aerial] part
地下サイロ	underground silo		
地下水	underground water	地上部成[生]長, 茎葉成[生]長, 茎の成[生]長	shoot growth
地下水位	ground water table [level]		
地下排水	underground drain		
地下部	subterranean [belowground, underground] part	致死量	lethal dose
		地図化, マッピング	mapping
		チゼルプラウ	chisel plow
地下部/地上部比	root/shoot ratio	地帯	zone

地代	land rent	昼間放牧	daytime grazing
地中茎	subcrown internode	中耕	intertillage
窒素	nitrogen, N	中日植物	day-neutral plant
窒素化合物	nitrogeneous compound	抽出法，抽出	extraction
		中心柱	stele
窒素過剰	nitrogen excess	中性デタージェント繊維	neutral detergent fiber, NDF
窒素きが【飢餓】	nitrogen starvation		
窒素源	nitrogen source	沖積扇状地，扇状地	alluvial fan
窒素固定	nitrogen fixation, N-fixation		
		沖積土	alluvial soil
窒素サイクル	nitrogen cycle	中絶，発育停止	abortion
窒素代謝	nitrogen metabolism	中層，中葉	middle lamella
窒素同化（作用）	nitrogen assimilation	抽たい【苔】	bolting
チトクローム酸化酵素	cytochrome oxidase	中毒	poisoning
		虫ばい【媒】	entomophily, insect pollination
千鳥植	zigzag planting		
地表	soil surface	中はい【胚】軸	mesocotyl
地表かんがい【灌漑】	surface irrigation	昼夜放牧	whole-day grazing
		中葉，中層	middle lamella
稚苗［幼苗］期	seedling stage	中肋，中央脈	midrib
地表植物	chamaephytes	中和	neutralization
地表［表面］流出［流去］	surface run-off	超遠心（分離）機	ultracentrifuge
ち【緻】密度	compactness	頂芽	apical [terminal] bud
地目	kind of land	鳥害	bird injury
着生節位	node position	頂芽優勢	apical dominance
着らい【蕾】枝	stem with a flower bud, flower stalk, stem bearing a flower bud	長期研究	long-term research
		長期予測	long-term prediction
		長期予報	long-range forecast
		長形ウイルス	elongated virus
チャンネル，水路	channel	徴候，標徴	sign
		長日植物	long-day plant
虫えい，虫こぶ	gall	長日処理	long-day treatment
中央値	median	長寿，寿命	longevity
中央脈，中肋	midrib	調整，調節，補正	adjustment
虫害	insect damage [injury]		
		頂生（の）	terminal
中間宿主	alternate [intermediate] host	調節，調整，補正	adjustment
		長草型	tall-grass type
中間生成物	intermediate product	頂端，頂部，せん【尖】部	apex
中間代謝	intermediary metabolism		

頂端分裂組織	apical meristem	土壌肥よく【沃】度	
超薄切片	ultrathin section		
頂部，頂端，せん【尖】部	apex	地力消耗植物	soil-exhaustive plant
		鎮　圧	compacting, firming, soil packing
重複遺伝子	duplicate gene		
重複[多重]感染	multiple infection		
		鎮圧輪	press wheel
頂部除去，断頭	decapitation	鎮圧ローラ	land roller, packer
長命種子	long-lived seed		
超優性	overdominance		つ
直　根	axial [tap] root		
直線回帰	linear regression	追加実験	additive [additional] experiment
直線性	linearity		
直腸検査	rectal palpation	追播，追播き	over-seeding [sowing], reseeding
直　播	direct seeding [sowing]		
直播栽培	direct seeding [sowing] culture	追播栽培	over-seeding culture
		追　肥	topdressing, additional manure
直立型	erect type		
直立茎，垂直茎	orthotropic stem [shoot], erect stem [shoot]	通気，ばっ【曝】気	aeration
		通気圧	aeration pressure
直立性	upright habit	通気性	air permeability
貯蔵，保存	preservation, conservation, storage	通気組織	pneumatic system
		通導組織	conductive tissue
		通年サイレージ給与	all-season silage feeding
貯蔵器官	reserve [storage] organ		
		通風乾燥	drying by drafting, forced-air drying
貯蔵根	storage root		
貯蔵性	storability	接　木	grafting, graftage
貯蔵草給与	storage feeding	繋飼い，繋牧	tether grazing, tethering
貯蔵組織	reserve [storage] tissue		
		詰込み，充てん【填】	filling
貯蔵物質	reserve substance		
ちょ【佇】立	standing	梅雨，雨期	rainy season
チラコイド	thylakoid	つ　る	vine
地理的隔離	geographical isolation	つる性(の)	viny
地理的分布	geographical distribution	ツンドラ	tundra
地理的変異	geographic variation		
治療，処理	treatment		
地力，	soil fertility		

て

T-R比	top-root ratio
TDN含量	TDN content
TDN収量	TDN yield
DNAマーカー	DNA marker
低栄養飼養	underfeeding
定温，恒温	constant [fixed] temperature
低温種子春化処理	low temperature seed vernalization
低温障害	low temperature injury
低温処理，冷処理	chilling
定温培養，インキュベーション	incubation
低温発芽性	low temperature germinability
低温要求度	chilling requirement
低下，衰弱，減退	decline
定芽	definite bud
抵抗性，耐性，抵抗力	resistance, tolerance
抵抗性検定	test for resistance
抵抗性品種	resistant variety
てい【蹄】耕法	hoof cultivation
抵抗力，抵抗性，耐性	resistance, tolerance
低コスト	low cost
低湿地土	meadow soil
低次［位］分げつ	lower order tiller, lower-nodal-position tiller
てい【蹄】傷，踏圧，踏付け	trampling, treading, compacting
てい【蹄】傷害	trampling [treading] damage
泥状物，スラリー	slurry
定植，植付け，栽植	planting
低水分サイレージ	low-moisture silage
ディスクハロー，砕土機	clod crusher, disk harrow
ディスクプラウ	disk plow
定性分析	qualitative analysis
泥炭，ピート	peat
泥炭土	peat soil
低地	lowland
定置式乾燥装置	stationary drier
低地草原	lowland grassland
定置放牧，固定放牧	fixed grazing [stocking], set stocking
定着	establishment
ディフェンス，防衛，防御	defence
低［かん【灌】］木	shrub, scrub
低マグネシウム血症	hypomagnesaemia
低養分環境	low-nutrient environment
定量分析	quantitative analysis
データベース	database
データロガー法	data logger method
デオキシリボ核酸	deoxyribonucleic acid, DNA
手刈り	hand cutting
適応	adaptation
適応性	adaptability
適応（的）反応［応答］	adaptive response
適応度	fitness
滴下法	drip culture
適期	optimum period
適合性，両立性，和合性	compatibility
適合度	goodness of fit

適正利用	proper use
滴定酸度	titratable acidity
摘葉処理, せん【剪】葉, 落葉	defoliation
テタニー	tetany
鉄	iron, Fe
テッダレーキ	tedder and rake
鉄砲ノズル	gun nozzle
デヒドロゲナーゼ, 脱水素酵素	dehydrogenase
デモグラフィ, 個体群統計学, 人口学	demography
テリトリ, 縄張り, 領域	territory
テレメータ	telemeter
田園アメニティ	rural amenity
添加	addition
添加剤 [物]	additive
転換畑	rotational upland field
天気	weather
電気泳動 (法)	electrophoresis
電気伝導率, 比電気伝導度	electric conductivity
電気牧柵	electric fence
転座, 転流, 移行	translocation
電子伝達系	electron transfer system
デンシトメトリ	densitometry
伝染	infection, transmission
転草	tedding
伝達, 伝搬	transmission
展着剤	spreader
天敵	natural enemy
伝導, 導通	conduction
伝統 (的) 農法	traditional agriculture [farming]
天然下種	natural seeding
天然供給 (量)	natural supply
天然更新	natural regeneration [reproduction]
天然苗, 自然実生	natural seedling
天然林	natural forest [stand]
点播	hill seeding [sowing], spaced planting
田畑輪換	paddy field and upland rotation, rotation between paddy and dry field
伝搬, 伝達	transmission
デンプン【澱粉】	starch
デンプン【澱粉】価	starch value, SV
デンプン【澱粉】粒	starch granule
転流, 転座, 移行	translocation

と

銅	copper, Cu
踏圧, てい【蹄】傷, 踏付け	trampling, treading, compacting
踏圧耐性, 耐踏圧性	trampling tolerance [resistance]
踏圧力	tread power
同位元素 [体], アイソトープ	isotope
同位 [イソ] 酵素, アイソザイム	isoenzyme, isozyme
同位種	ecological equivalent
同位体 [元素], アイソトープ	isotope
頭花	flower head
豆果	pod, legume
透過	transmittance
同化 (作用)	assimilation
同化器官	assimilatory organ

同化産物	assimilation product	登熟期	ripening stage
透過性, 浸透性, 透過率	permeability	登熟期間	ripening period
		凍上	frost heaving, heaving
同化デンプン【澱粉】	assimilatory starch	頭数調整放牧	put-and-take stocking
		透析（法）	dialysis
同化箱法	assimilation chamber method [technique]	動態	dynamics
		糖タンパク【蛋白】質	glycoprotein
透過率, 透過性, 浸透性	permeability	等張液	isotonic solution
道管《植物》	vessel	導通, 伝導	conduction
冬季放牧	winter (season) grazing	同定, 識別	identification
		等電点	isoelectric point
統計遺伝学	statistical genetics	導入	introduction
統計学	statistics	導入育種	introduction breeding
統計学的	statistic, statistical	導入遺伝資源	introduced genetic resources
同系交配, 近親交配 [近交]	inbreeding	導入作物	introduced crop
		導入（草）種	introduced species
統計処理	statistical treatment	導入品種	introduced variety
同型 [ホモ] 接合体	homozygote	当年茎	annual [current-year] tiller [stem]
統計調査	statistical survey		
同系繁殖, 内交配	inbreeding	当年生, 一年生（の）	annual
統計分析	statistical analysis	同伴イネ科牧草	companion grass
凍結	freeze	同伴作物	companion crop
凍（結）害	freezing damage [injury]	同伴種	companion species
		逃避反応	escape response
凍結真空乾燥	freeze drying, lyophilization	豆腐かす【粕】	tofu cake
		倒伏	lodging
動向, 傾向	trend	倒伏抵抗性, 耐倒伏性	lodging resistance
等高線栽培	contour cropping		
等高帯栽培	contour strip cropping	倒伏防止	lodging control
塔サイロ, タワーサイロ	tower silo	動物生態学	animal ecology
		動物相	fauna
投資	investment	動物多様性	faunal diversity
糖脂質	glucolipid	動物福祉	animal welfare
糖（質）	sugar	糖蜜	molasses
同質倍数体	autopolyploid, autoploid	冬眠	hibernation, wintering
		トウモロコシかん【稈】	corn stover
同質四倍体	autotetraploid		
同質六倍体	autohexaploid		
登熟	grain filling, ripening		

トウモロコシ収穫機, コーンハーベスタ	corn harvester	土壌改良	soil amendment [amelioration, improvement]
トウモロコシ点ま【蒔】き機, コーンプランタ	corn planter	土壌改良剤	soil amendment matter, soil conditioner
		土壌化学	soil chemistry
		土壌学	soil science, pedology
トウモロコシ粒	shelled corn	土壌型	soil type
投与量, 服用量, 線量	dose	土壌くん【燻】蒸	soil fumigation
動力取出し軸	power take-off shaft, PTO	土壌構造	soil structure
		土壌コロイド	soil colloid
糖類	saccharide	土壌殺菌	soil disinfection [sterilization]
登録	registration		
登録種子	registered seed	土壌酸性	soil acidity
登録飼料	registered feed	土壌三相	three phases of soil
登録品種	registered variety	土壌資源	edaphic resources
土塊	clod	土壌侵食	soil erosion
時なし性《周年出穂性》	non-seasonal	土壌診断	soil diagnosis
		土壌深度	soil depth
特性, 形質	character	土壌図	soil map
特性(の)	characteristic	土壌水	soil water
毒性	virulence, toxicity	土壌水分	soil moisture
特性検定	test of specific character	土壌水分含量	soil moisture [water] content
毒素	toxin	土壌生産力	soil productivity
特定組合せ能力	specific combining ability	土壌生物	soil organism
		土壌接種	soil inoculation
毒物	toxicant	土壌層位	soil horizon
独立栄養生物, 無機栄養生物, 独立栄養体	autotroph	土壌退化	soil degradation
		土壌断面	soil profile
		土壌団粒	soil aggregate
独立栄養体, 無機栄養生物, 独立栄養生物	autotroph	土壌調査	soil survey
		土壌伝染病	soil-borne disease
		土壌統	soil series
トコフェロール, ビタミンE	tocopherol, vitamin E	土壌動物	soil animal
		土壌動物相	soil fauna
土質	soil property	土壌反応	soil reaction
土壌	soil	土壌微生物	soil microorganism
土壌汚染	soil pollution	土壌微生物学	soil microbiology
土壌害虫類	soil pest		

土壌肥よく【沃】 度, 地力	soil fertility	回路, クエン 酸回路	[TCA] cycle
土壌肥料学	science of soil and plant nutrition	トリッピング	tripping
		ドリル, 条播機	drill
土壌物理学	soil physics	ドリル播き	drilling
土壌物理性	soil physical property	トルエン蒸溜法	toluene distillation method
土壌分類	soil classification		
土壌 pH	soil pH	トレーサー法	tracer technique
土壌保全	soil conservation	トレードオフ	trade-off
土壌有機物	soil organic matter	トレーラ	trailer
度数曲線	frequency curve	トレンチサイロ	trench silo
度数 [頻度] 分布	frequency distribution	ドロマイト	dolomite
土 性	soil texture		な
土層 [心土] 改良	subsoil improvement	内えい【穎】	palea
土地回復 [修復]	land restoration	内果皮	endocarp
土地改良	land improvement	内交配, 同系繁殖	inbreeding
土地交換分合	consolidation of land		
土地生産性	land productivity	内生休眠	innate dormancy
土地造成, 干拓	land reclamation	内生菌, エンド ファイト	endophyte
土地評価	land value		
土地分類	land classification	内生植物 ホルモン	endogenous plant hormone
徒 長	spindly growth		
土地利用	land utilization [use]	内生 (の)	endogenous
凸状草の	hummock	内毒素, 菌体内 毒素, エンド トキシン	endotoxin
突然変異	mutation		
突然変異体, 変異体, 変異株	mutant		
		内 皮	endodermis
		内分泌かく乱 化学物質	endocrine disruptors
トップ アンローダ	top unloader		
トップグレー ジング, 上層採食	top grazing	ナイロンバッグ 法	nylon bag method
		中生 (の)	medium-maturing
		ナタネかす【粕】	rapeseed meal
トップ交雑	top-cross	なたハロー	colter harrow
ドナー, 供与体	donor	夏枯れ	summer depression
土 膜	crust	夏 作	summer cropping
止 葉	flag [boot] leaf	夏作飼料作物	summer forage crop
トラクタ	tractor	夏胞子	urediniospore
ドラムモーア	rotary drum mower	夏山冬里方式	winter housing system
トリカルボン酸	tricarboxylic acid	ナトリウム	sodium, Na

生ふん【糞】, 新鮮ふん【糞】	fresh feces
鉛	lead, Pb
縄張り, 領域, テリトリ	territory
軟毛（でおおわれている）	pubescence, pubescent

に

二価染色体	bivalent chromosome
二期作	double cropping
肉牛生産 [畜産]	beef production
肉食性昆虫	carnivorous insect
肉（用）牛	beef cattle
肉用牛経営	beef cattle farming
肉用牛繁殖経営	cow-calf farming
肉用牛肥育経営	cattle fattening farming
二項分布	binomi(n)al distribution
ニコチンアマイドアデニンジヌクレオチド	nicotinamide adenine dinucleotide, NAD
ニコチンアマイドアデニンジヌクレオチドリン酸	nicotinamide adenine dinucleotide phosphate, NADP
二酸化炭素, 炭酸ガス	carbon dioxide [CO_2]
二シーズン放牧	two-season grazing
二次曲線	quadratic curve
二次根	secondary root
二次細胞壁	secondary cell wall
二次生産	secondary production
二次成 [生] 長	secondary growth
二次遷移	secondary succession
二次代謝物	secondary metabolite
二次分げつ	secondary tiller
二次林	secondary forest
日エネルギー	daily energy
要求量	requirement
日較差	diurnal range
日増体量	daily gain
日内変動, 日変化	diurnal variation
日変化, 日内変動	diurnal variation
日 乾	curing
日間リズム	diurnal rhythm
ニッケル	nickel, Ni
日射反射率	albedo
日射（量）	(amount of) solar radiation
日周行動	daily behavior
日照計	sunshine recorder
日照時間	duration of sunshine
ニッチ, ニッチェ	niche
日 長	day length, photoperiod
日長調節	photoperiodic control
日長 [光周] 反応	photoperiodic response
二年生（の）	biennial
二倍体（の）, 複相体（の）	diploid
2番芽	second flush
2番草	second flush
2番草放牧	aftermath grazing
日本短角種	Japanese Shorthorn
二面交雑 [交配], 総当たり交雑 [交配]	diallel crosses
乳 酸	lactic acid
乳酸菌	lactic acid bacteria
乳酸発酵	lactic fermentation
乳 質	milk quality
乳飼比	rate of feed cost to milk sold
乳脂肪	milk fat
乳汁分泌, 泌乳, 授乳	lactation

乳熟	milk-ripe	ネザサ型草地, アズマネザサ型草地	Pleioblastus-type grassland
乳熟期	milk-ripe stage		
乳脂率	butter fat [milk fat] percentage	熱収支	heat budget [balance]
乳成分	milk composition, milk ingredient	熱帯	tropical zone, tropics
		熱帯草地	tropical grassland
乳せん【腺】	mammary gland	熱帯の, 暖地型(の)	tropical
乳せん【腺】炎, 乳房炎	mastitis		
		熱電対	thermocouple
乳糖	lactose	熱伝導	heat transfer
乳頭	teat	熱風乾燥, 加熱乾燥	heated-air drying
乳房	udder		
乳房炎, 乳せん【腺】炎	mastitis	熱風乾燥機, 火力乾燥機	heated-air dryer
入牧	turning out to pasture, setting up the grazing	熱放射	thermal radiation
		熱量	heat [thermal] capacity
乳(用)牛	dairy cattle	根の分泌物	root exudate
乳用種	dairy breed	根ばり	root spread
乳量	milk yield	根雪期間	continuous snow cover duration
尿	urine		
尿散布機	urine spreader	年次相関	year-to-year [yearly] correlation
尿素	urea		
尿やけ	urine scorch	ねん【稔】実	ripening
妊娠	gestation, pregnancy	ねん【稔】実歩合[率]	percentage of ripening
ぬ			
		粘性, 粘度	viscosity
		ねん【稔】性, 繁殖性, 受精能	fertility
ぬか類	brans		
沼, ラグーン	lagoon		
沼地, 湿地	marsh	粘ちょう【凋】度, コンシステンシー, 密度	consistency
ね			
		粘土, しょく【埴】土	clay
根	root		
根株《樹の》, スタンプ	stump	粘度, 粘性	viscosity
		燃料	fuel
根ぐされ ネクローシス, え死	root rot necrosis	年輪	annual ring
		年令[齢]	age
根こぶ	club root		
根こぶ線虫	root knot nematode		

の

農学	agronomy
農機具	farming machinery
農業改良普及員	agricultural extention worker [officer]
農業機械	agricultural machinery, farm machinery
農業機械化	farm mechanization
農業気象学	agricultural meteorology
農業経営費	expenditure for agriculture, farm management cost, cost of farming
農業［実用，経済］形質	economic [agronomic] character [trait]
農業構造	agricultural structure
農業集落，農村	farm village, rural community
農業所得	agricultural income
農業生産	agricultural production
農業粗収益	gross agricultural income
農業廃棄物	agricultural wastes
農具舎	barn
濃厚飼料	concentrate (feed)
濃縮（物）	concentration
農場管理	farm management
のう状樹枝状菌根，VA菌根	vesicular arbuscular mycorrhizas
農村，農業集落	farm village, rural community
農村地域	rural area
濃度《溶液の》	concentration
能動的吸収，積極的吸収	active absorption
能動輸送	active transport
濃度障害，塩類障害	salt damage [injury]
農薬	pesticide, agricultural chemical
能力［性能］検査	performance test
のぎ，ぼう【芒】	awn, arista
野積み，スタック	stack
野ねずみ，げっ歯類	rodent
野火	fire
乗換え，交叉，交差	crossing over
法面	slope face, slope

は

葉	leaf
葉《集合的》，茎葉	foliage
パーオキシダーゼ	peroxidase
バークたい【堆】肥，樹皮たい【堆】肥	bark compost
パーティクルガン	particle gun
ハードニング，硬化	hardening
バーナリゼーション，春化	vernalization
ハーベスタ，収穫機	harvester
バーミキュライト	vermiculite
バーンクリーナ	barn cleaner
はい【胚】，胎仔	embryo
梅雨，雨期	rainy season
排液法，排水，排のう【膿】法	drainage

は [65]

バイオアッセイ, 生物（学的）検定	bioassay
バイオーム 《生物群系》	biome
バイオスフェア, 生物圏	biosphere
バイオテクノロジ	biotechnology
バイオテレメトリ	biotelemetry
バイオトロン	biotron
バイオマス, 生物（体）量	biomass
バイオマス生産	biomass production
バイオリアクタ	bioreactor
バイオリメディエーション, 生物環境修復	bioremediation
ばい【媒】介昆虫	insect vector
排　気	air drainage
廃棄物, ふん【糞】尿	waste
配偶子	gamete
はい【胚】形成	embryogenesis
配合禁忌, 配合変化	incompatibility
配合原料	feed ingredient
配合飼料	formula feed
配合肥料	blended [mixed] fertilizer, composed manure
配合変化, 配合禁忌	incompatibility
はい【胚】軸, 下はい【胚】軸	hypocotyl
はい【胚】軸培養	hypocotyl culture
はい【胚】珠	ovule
排出, 置換	displacement
排出, 排泄	excretion
はい【胚】珠培養	ovule culture
排除実験	exclusion experiment
排水, 排液法, 排のう【膿】法	drainage
廃　水	waste water
倍数化	polyploidization
倍数性	polyploidy
倍数性育種	polyploidy breeding
倍数体, 多数性	polyploid
排泄, 排出	excretion
排泄物	excreta, excretion
排泄量	amount of excreta
培地, 培養液	culture medium
バイト, 食いちぎり, か【噛】み	bite
配糖体, グルコシド	glucoside
バイトサイズ	bite size
バイト重	bite weight
バイト深	bite depth
バイト速度	bite rate, biting rate
バイト面積	bite area
バイト容積	bite volume
バイト量	bite mass
はい【胚】乳	endosperm
排　尿	urination
排のう【膿】法, 排水, 排液法	drainage
はい【胚】培養	embryo culture
バイパスタンパク【蛋白】質	by-pass protein
はい【胚】盤, 小板, 盤状体	scutellum
パイプダスタ	pipe duster
排ふん【糞】	defecation

排ふん【糞】過繁地，不食過繁地，不食地	dung patch	播種法	seeding [sowing] method
		播種密度，播種量	seeding [sowing] rate
配分，分配	allocation, partition	播種量，播種密度	seeding [sowing] rate
排ふん【糞】地	area covered by dung, dung-deposited place [area]	パスチャハロー	pasture harrow
		破生通気組織	lysigenous aerenchyma
培養	culture	畑作	field crop cultivation
培養液，培地	culture medium	畑作物	upland [field] crop
バガス《さとうきびの搾りかす【粕】》	bagasse	畑(地)，ほ【圃】場	field
吐戻し，逆流，弁閉鎖不全	regurgitation	畑［台地］土壌	upland soil
		波長	wavelength
白化，黄変，クロロシス	chlorosis	発育，発生	development
		発育相	developmental phase
麦芽糖，マルトース	maltose	発育阻止，すくみ，わい【矮】化	stunt
白色種，アルビノ，白子	albino		
		発育停止，中絶	abortion
薄層クロマトグラフィ	thinlayer chromatography, TLC	発芽	germination
		発芽試験	germination test
		発芽習性	germination habit
はく皮処理	peeling treatment	発芽勢	germination ability, viability of seed
バケットエレベータ	bucket elevator		
		発芽促進，催芽	hastening of germination
バケットドーザ	bucket dozer		
は【耙】耕，ハローイング	harrowing	発芽遅延	delayed germination
		発芽日数	number of days for germination
破砕	crush		
播種，たねまき	seeding, sowing	発芽率	germination percentage, percentage of germination
播種機	seeding [sowing] machine, seeder		
播種(時)期	seeding [sowing] time		
		ばっ【曝】気，通気	aeration
播種深(度)	seeding [sowing] depth	バッグサイロ	bag silo
		発酵	fermentation
播種床	seeding [seed, sowing] bed	発酵乾燥，たい【堆】肥化，コンポスト化	composting
播種床造成	seed bed preparation		
播種日	seeding [sowing] date		

発酵品質	fermentative quality	半乾燥地	semi-arid region
発根	rooting	はん【汎】骨	panmyelopathia
抜根	grubbing	ずいろう	
抜根機	stump puller	【髄癆】	
発根節	rooted node	伴細胞	companion cell
発情回帰	recurrence [return] of estrous	反作用，反応	reaction
		半自然草地	semi-natural grassland
発情（期）	estrous	［草原］	
発情周期, 性周期	estrous cycle	反射（率）	reflectance, reflectivity
発生，発育	development	晩熟	late maturation
発生予察	forecasting of occurrence	晩熟性	late maturity
		繁殖《動物》, 生殖《動物》	reproduction, breeding
発達，開発	development		
パッチ，過繁地	patch	繁殖《動物》, 育種	breeding
パッチ状	patchiness		
パッチ状採食	patch [spot] grazing	繁殖型	migrule form
パッチ状分布	patchy distribution	繁殖期，生殖成［生］長期	reproductive stage
パッチモデル	patch model		
はつ土板プラウ	moldboard plow	繁殖器官	reproductive part
バッファ	buffer	繁殖季節	breeding season
バッファロ， 水牛	buffalo	繁殖牛	breeding cattle, reproductive cattle
花	flower	繁殖茎	reproductive tiller [stem]
放飼式牛舎， ルースバーン	free barn, loose barn		
		繁殖効率, 繁殖投資	reproductive effort
葉の，葉状の， 葉質の	foliar	繁殖障害	reproductive difficulties
葉の寿命	leaf longevity		
早刈り	early harvesting	繁殖性，	fertility
早播き， 早期播種	early seeding [sowing]	ねん【稔】性， 受精能	
バリウム	barium, Ba	繁殖投資, 繁殖効率	reproductive effort
春肥	spring dressing		
春播き	spring seeding [sowing]	繁殖肥育一貫経営	cow-calf finishing operation
春播性	spring habit	繁殖分配	reproductive allocation
ハロー	harrow		
ハローイング, は【耙】耕	harrowing	繁殖率［力］, 種子生産量	fecundity, reproductivity
バンカーサイロ	bunker silo	繁殖力［率］, 種子生産量	fecundity, reproductivity
半乾燥（の）	semi-arid		

反すう【芻】	rumination
反すう【芻】胃, 第一胃, ルーメン	rumen
反すう【芻】胃, 第一・二胃	reticulorumen
反すう【芻】家畜［動物］, 反すう【芻】の	ruminant
反すう【芻】行動	rumination behavior
反すう【芻】時間	rumination time
半数体, 単相体, 一倍体	haploid
半数致死量	lethal does 50, LD$_{50}$
反すう【芻】動物［家畜］, 反すう【芻】の	ruminant
反すう【芻】の, 反すう【芻】動物［家畜］	ruminant
晩生品種	late variety
晩霜	late frost
反草機	swath turner
半地中植物	hemicryptohytes
反転	turning
反転飼養試験法	switch-back feeding experiment
反転法	change-over design
半透膜	semipermeable membrane
ハンドリング, 操作	operation, handling
反応, 反作用	reaction
晩播, 遅播き	late seeding [sowing]
判別関数	discriminant function
判別寄主	differential host
はん【斑】紋	mottle
はん【汎】用トラクタ	general purpose tractor

ひ

PCR法	PCR [polymerase chain reaction] method
pF-水分曲線	pF-soil moisture curve
pF値	pF-value
肥育	fattening
肥育牛	fattening [feeding, store] cattle
肥育用素牛	feeder [stock] cattle
ヒース（の茂る荒れ地）	heathland
ピート, 泥炭	peat
ビートパルプ	beet pulp
ビールかす【粕】	brewer's grain
火入れ	burning
火入れした土地	burned-over land
ひ【庇】蔭	shelter, shade
ひ【庇】陰栽培	shading culture
ひ【庇】陰牧草	shading pasture plant
ひ【庇】陰林	shelter woods
ビオトープ	biotope
被害	damage
比較品種, 対照品種	check variety
比活性, 比放射能	specific activity
光エネルギー	light energy
光エネルギー転換効率	efficiency for light energy conversion
光強度, 照度	light intensity, illuminance
光形態形成	photomorphogenesis
光呼吸	photorespiration
光透過率	light transmission
光発芽, 明発芽	light germination
光発芽種子	dark-inhibited seed, light germinator, light sensitive seed

ひ [69]

光分解	photolysis, photodecomposition
光飽和	light saturation
光補償点	light compensation point
微気候	microclimate
引き抜き	pulling
ひき割り, 荒粉	meal
低刈り	low-level cutting
ひげ根, 繊維根	fibrous root
肥効	effect of fertilizer, manuring effect, fertilizer efficiency
肥効試験	fertilizer response test
非構造性炭水化物	non-structural carbohydrate, NSC
肥効調節型肥料	controlled release fertilizer
ひこばえ, 台芽	sucker
微細環境, 微小環境	microenvironment, microsite
微細構造	ultrastructure
被子植物	angiosperm
皮質, 皮層	cortex
微砂	silt
比重, 真比重	specific gravity
比重選	selection by specific gravity
微小環境, 微細環境	microenvironment, microsite
微小体, マイクロボディ	microbody
被食	herbage consumption
被食耐性	defoliation [browse] tolerance
比色分析	colorimetric analysis
微生物	microorganism
微生物相, ミクロフローラ	microflora
非生物的環境	abiotic factor
皮層, 皮質	cortex
肥大	thickening
肥大成[生]長	thickening growth
ビタミン	vitamin
ビタミンA_1, レチノール	vitamin A_1, retinol
ビタミンB_1, チアミン	vitamin B_1, thiamin
ビタミンB_2, リボフラビン	vitamin B_2, riboflavin
ビタミンB_6, ピリドキシン	vitamin B_6, pyridoxin
ビタミンC, アスコルビン酸	vitamin C, ascorbic acid
ビタミンD, カルシフェロール	vitamin D, calciferol
ビタミンE, トコフェロール	vitamin E, tocopherol
非タンパク【蛋白】態窒素	non-protein nitrogen, NPN
備蓄用牧区	autumn saved pasture, ASP
微地形	microtopography
ピックアップユニット	pick up unit
ピックアップワゴン	pick up [self-loading] forage wagon
ビッグラウンドベーラ, ビッグロールベーラ	big round baler, big roll baler
ビッグロールベーラ, ビッグラウンドベーラ	big round baler, big roll baler
羊, めん【緬】羊	sheep
必須アミノ酸	essential amino acid

必須元素, 必須要素	essential element	比活性	
		被毛	hair
必須脂肪酸	essential fatty acid	非毛管孔げき【隙】	non-capillary pore
必須要素, 必須元素	essential element		
		百万分率	parts per million, ppm
比電気伝導度, 電気伝導率	electric conductivity	病因, 病原（体）	causal agent, pathogen
比伝導度	specific conductivity	評　価	assessment, evaluation
被　度	coverage		
微動遺伝子, ポリジーン, 多遺伝子	polygene	ひょう【雹】害	hail injury
		病　害	disease
		病害抵抗性, 耐病性	disease resistance [tolerance]
避難林	shelter trees		
泌乳, 授乳, 乳汁分泌	lactation	病害ばい【媒】介者	vector
泌乳期間, 授乳期	lactation period	病害発生予察	disease forecasting
		病原（体）, 病因	causal agent, pathogen
泌乳牛, 搾乳牛	milking [lactating, dairy] cow		
		表現型 [形]	phenotype
泌乳曲線	lactation curve	病原型, 寄生型	pathotype
泌乳試験	lactation experiment	表現型可塑 [変] 性,	phenotypic plasticity
泌乳能力	dairy [milk] performance		
		病原菌	pathogenic fungus
泌乳量	milk producing yield	病原体	causal organism
肥培管理	manuring practice	病原力 [性]	pathogenicity, virulence
非発根節	no-rooted node		
非病原性, 弱毒性	avirulence	標　高	altitude, elevation
		標識法	marking method
皮膚温	skin temperature	標準誤差	standard error
被覆, コーティング	coating	標準品種	standard variety
		標準偏差	standard deviation
被覆, マルチ	mulch	苗条, 若枝, 茎 (部)	shoot
被覆栽培, マルチ栽培	mulch culture, mulching cultivation		
		表　層	surface horizon
		表層施肥	surface (layer) application
被覆作物	cover crop		
被覆度	cover degree	標徴, 徴候	sign
非分解性タンパク【蛋白】質	undegradable protein	病　徴	symptom
		病徴隠ぺい, 不顕性の, しゃへい【遮蔽】	masking
非平衡モデル	non-equilibrium model		
比放射能,	specific activity		

標徴種	characteristic species			micronutrient,
表　土	topsoil			minor element,
病はん【斑】,	lesion			trace element
病変		ピルビン酸	pyruvic acid	
表　皮	epidermis	ピレノイド,	pyrenoid	
表皮細胞	epidermal cell	核様体		
病変,	lesion	広うね【畦】	broad ridge	
病はん【斑】		ひろがり型	spreading type	
標本, 試料,	sample	ピロプラズマ病	piroplasmosis	
検体		貧栄養の,	oligotrophic	
標本抽出,	sampling	栄養不足の		
試料採取		[不良の]		
標本平均	sample mean	貧血（症）	anemia	
表面散布	broadcasting of	品質, 質	quality	
	fertilizer	品　種	variety	
表面積	surface area	品種, 栽培品種	cultivar	
比葉面積	specific leaf area,	品種証明	varietal certification	
	SLA	品種特性	varietal characteristics	
表面播種	surface sowing	品種比較試験	variety test	
表面［地表］	surface run-off	品種保存	variety preservation	
流出［流去］		頻度, 回数	frequency	
苗　令	seedling age	頻度依存	frequency dependent	
肥よく【沃】地	fertile land	頻度［度数］	frequency distribution	
肥よく【沃】度,	fertility	分布		
肥料分				
ピリドキシン,	pyridoxin, vitamin B$_6$		ふ	
ビタミン B$_6$				
肥　料	fertilizer, manure	ファームワゴン	farm wagon	
微量元素［要素］	microelement,	ファイトクロー	phytochrome	
	micronutrient,	ム, フィト		
	minor element,	クロム, フィ		
	trace element	トクローム		
肥料三要素	three primary	ファイトトロン,	phytotron	
	nutrients, three	人工気象室		
	major nutrients,	《植物用》		
	N,P,K elements	ファイトプラズ	phytoplasma	
肥料試験	fertilizer trial	マ, マイコ		
肥料成分	fertilizer nutrient	プラズマ様		
肥料分, 肥よく	fertility	微生物		
【沃】度		部　位	part	
微量ミネラル	trace mineral	フィーディング	feeding station	
微量要素［元素］	microelement,	ステーション		

フィードバック	feedback	フォレージ ブロワ	forage blower
VA菌根, のう 状樹枝状菌根	vesicular arbuscular mycorrhizas	フォレージ ワゴン	forage wagon
フィステル, ろう孔 [管]	fistula	富　化	enrichment
フィトクロム, フィトクロー ム, ファイト クローム	phytochrome	不可給態要素	non-available [unavailable] element
		ふ【賦】活 (化), 活性化,	activation
フィンガホイー ルレーキ	finger-wheel rake	ふ【賦】活剤,	activator
風　化	weathering	活性化物質,	
風　害	wind damage	活性化因子	
風乾重	air-dry weight	不活性化, 失活	inactivation
風乾土	air-dry soil	不完全菌	imperfect fungi
風乾物	air-dry matter	不完全時代,	imperfect state
風　食	wind erosion	無性時代	
風　選	winnowing, wind selection	吹上カッター	ensilage cutter
		不規則	random distribution
風　速	wind velocity	[ランダム,	
風速計	anemometer	機会的] 分布	
風ばい【媒】	anemophily	普　及	extention
風　味	flavor	不吸湿性種子	impermeable seed
フェザーミール, 羽毛粉	feather meal	不均一 [質] 性, 異質性, 異種性	heterogeneity
フェノール酸	phenolic acid, phenol		
フェノール性 物質	phenolic substance [compound]	複合経営	diversified [multiple] farming
フェノロジー, 生物季節学	phenology	副産物	by-product
		複式遺伝子型	duplex genotype
フォーレージ テスト	forage test	複相体 (の), 二倍体 (の)	diploid
フォスフォエ ノールピルビ ン酸	phosphoenol pyruvic acid, PEP	複相胞子生殖	diplospory
		複対立遺伝子	multiple alleles
		覆　土	covering with soil, molding
フォスフォリラ ーゼ, 過リン 酸分解酵素	phosphorylase	複二倍体	amphidiploid
		複　葉	compound leaf
フォッゲージ, 霜枯れ草	foggage	服用量, 投与量, 線量	dose
フォレージ ハーベスタ	forage harvester		

ふ [73]

不顕性の，病徴隠ぺい，しゃへい【遮蔽】	masking	ふつ【沸】石，ゼオライト	zeolite
		フッ素	fluorine, F
不耕起	no-tillage	物理的	physical
不耕起栽培	non-tilled cropping, plowless farming, zerotillage	不定芽	adventitious bud
		不定根	adventitious root
		不適合性，不親和性，不和合性	incompatibility
不耕起草地造成	non-tilled establishment, unplowed grassland establishment	不透性層	impermeable layer
		ブドウ糖，グルコース	glucose
不耕起追播	unplowed reseeding	歩留り，回収（率），回復	recovery
不耕起ドリル播き	drillage	不ねん【稔】種子	sterile seed
不耕起播き	sod-seeding [sowing]	不ねん【稔】性	infertility, sterility
不食過繁地，不食地，排ふん【糞】過繁地	dung patch	不ねん【稔】(の)	sterile
		腐敗	putrefaction
		部分開花	partial blooming
腐植酸	humic acid	部分耕うん【耘】	partial tillage
腐植（質）	humus	部分耕起	partial plowing
不食草	unparatable [wolf] grass [plant]	部分耕栽培	partial tillage culture
		部分優性	partial dominance
不食地，不食過繁地，排ふん【糞】過繁地	dung patch	不飽和脂肪酸	unsaturated fatty acid
		踏付け，てい【蹄】傷，踏圧	trampling, treading, compacting
不親和性，不和合性，不適合性	incompatibility	不毛地	infertile land
		冬枯れ	winter killing
ふすま《表皮》	wheat bran	冬穀物	winter cereal
腐生者[菌]，雑菌	saprophyte	冬作	winter cropping
		冬胞子	teleutospore
豚	pig	冬芽	winter bud
不断給餌	adlibitum feeding	不溶性炭水化物	insoluble carbohydrate
物質	substance, material, matter	フラクトース，果糖	fructose
物質循環	material [matter] cycle, circulation of material	フラクトサン，フルクトサン	fructosan
		プラスチックフィルム	plastic film
物質生産	matter production		
ブッシュカッタ，刈払い機	bush cutter		

プラスチド, 色素体	plastid
フラックス	flux
フラボノイド	flavonoid
ブリックス糖度	Blix
フリントコーン	flint corn
フルクタン	fructan
フルクトサン, フラクトサン	fructosan
フレーク	flake
プレーリー 《北米中部の内陸草原》	prairie
フレール型フォレージハーベスタ	flail type forage harvester
フレールモーア	flail mower
フローラ, 植物相	flora
プロテアーゼ, タンパク【蛋白】質分解酵素	protease
プロトゾア, 原生動物, 原虫	protozoa
プロトプラスト, 原形質体	protoplast
フロリゲン, 花成素	florigen
プロリン	proline
ブロワ, 送風機	blower
フロントローダ	front-end loader
不和合性, 不親和性, 不適合性	incompatibility
ふん【糞】, ふん【糞】便, 大便	feces
分域電気泳動, ゾーン電気泳動	zone electrophoresis
粉衣法	dust coating method
分化	differentiation
ふん【糞】塊	dung pat
分解	decomposition
分解者	decomposer
分解性タンパク【蛋白】質	degradable protein
分化型	forma specialis, f. sp. [pl. formae speciales]
分画, 分取	fractionation
分割区試験法	split plot experiment
分げつ	tiller
分げつ芽	tiller bud
分げつ期	tillering stage
分げつ構成	tiller composition
分げつ習性	tillering habit
分げつ数	tiller number
分げつ盛期	active tillering stage
分げつ節	tillering node
分げつ密度	tiller density
分光光度計	spectrophotometer
分光分析	spectrophotometric analysis
粉砕, 圧搾	crush
粉砕, 研磨	grinding
粉砕機	crushing mill
分散	variance
分散, 散布	dispersion, dispersal
分散図	dispersion map
分散比	variance ratio
分散分析	analysis of variance, ANOVA
分枝	branch(ing)
分施	split application [dressing]
分枝型	branching form [type]
分取, 分画	fractionation
分集団	sub-population
分子量	molecular weight
分生子	conidium [pl. conidia]

分生子層	acervulus [pl. acervuli]	平均温度	mean temperature
分生子柄	conidiophore	平均（値）	mean
フン虫《食ふん【糞】性コガネムシ》	dung beetle	平均平方	mean square
		ヘイクラッシャ, 乾草破砕機	hay crusher
ふん【糞】尿, 廃棄物	waste	平原	plain
		平衡	equilibrium
ふん【糞】尿還元	return of animal excreta	平衡モデル	equilibrium model
		ヘイコンディショナ, 乾草圧砕機	hay conditioner
ふん【糞】尿処理	manure management	閉鎖花受精	cleistogamy
分配, 配分	allocation, partition	柄子殻	pycnidium [pl. pycnidia]
分泌	secretion		
分泌物	exudate	ヘイスタッカ, 乾草積上機	hay stacker
分布	distribution		
分布型	distribution type	ヘイテッダ, 乾草反転機	hay tedder
分布地	distribution area		
分布様式	pattern of spatial distribution, mode of distribution	平年作	normal crop
		ヘイフォーク	hay fork, pitch fork
		ヘイプレス, 乾草圧縮機	hay press, stationary hay baler
ふん【糞】便, ふん【糞】, 大便	feces	ヘイベーラ, 乾草こん【梱】包機	hay baler
分娩	calving, parturition		
噴霧機, 散布器	sprayer	柄胞子	pycnidiospore
噴霧接種	spray inoculation	ヘイリフト, 乾草揚げ機	hay lift
分離	segregation		
分離株	isolate	ヘイレーキ	hay rake
分離の法則	law of segregation	ヘイレージ	haylage
分類	classification	ヘイローダ, 乾草積込機	hay loader
分類学	taxonomy		
分裂	division	ベール, こん【梱】包	bale
分裂前期, 前期	prophase		
分裂組織, 成[生]長点	meristem	ベールスローワ	bale thrower
		ベールローダ	bale loader
		ベールワゴン	bale wagon
へ		ベクター	vector
		ペクチン	pectin
β-カロチン	β-carotene	ペクチン酸	pectic acid
ヘイエレベータ	hay elevator	ヘテローシス, 雑種強勢	heterosis, hybrid vigor
ヘイキューブ	hay cube		

ヘテロ[異型, 異種]接合体	heterozygote	穂状花序	
		保安林	protection forest
ヘミセルロース	hemicellulose	ホイールカッタ	wheel cutter
ヘモグロビン, 血色素	hemoglobin	ポイントコドラート法	point quadrat method
ヘリウム	helium, He	ほう【苞】, ほう【苞】葉	bract
ペレット, 固形飼料	pellet	ほう【芒】, のぎ	awn, arista
変異	variation	包えい【穎】, えい【穎】	glume
変異株, 突然変異体, 変異体	mutant	防衛, 防御, ディフェンス	defence
変異[変動]係数	coefficient of variation, CV	防えき【疫】, 検えき【疫】	quarantine
変異性, 可変性	variability	ほう【萌】芽, 出芽	budding, emergence
変異体, 突然変異体, 変異株	mutant	ほう【萌】芽	sprout
		妨害極相	disclimax
変温	alternating temperature	訪花昆虫	flower-visiting insect
		防火帯	firebreak
変温性の	poikilothermal	防かび剤	antimold
偏回帰	partial regression	ほう【萌】芽林《林業》	coppice [sprout] forest
変換効率	conversion efficiency		
偏向, 偏差	deviation	放棄地	abandoned [uncultivated] land
偏差, 偏向	deviation		
変性	denaturation	防御, 防衛, ディフェンス	defence
変性, 退化, 逆行変性	degeneration		
		方形区, コドラート, 方形コドラート	quadrat
変遷	changes		
偏相関	partial correlation		
変動, 異形	variation	方形区法, 区画法, 枠法	quadrat method
変動因	source of variation		
弁閉鎖不全, 吐戻し, 逆流	regurgitation	方形コドラート, コドラート, 方形区	quadrat
変量	variate		
		方形ハロー	square harrow
ほ		ほうこう【彷徨】変異	fluctuation
穂	ear, head	豊作, 上作	abundant [good] harvest
穂, 円すい【錐】花序	panicle		
		ホウ素	boron, B
穂, 花穂,	spike	胞子, 芽胞	spore

ほ　[77]

胞子形成	sporulation	放牧時間	grazing time
放飼効果	release effect	放牧施設	grazing facility
放射, 放射線,	radiation	放牧周期	grazing cycle
照射		放牧草	grazing grass
放射維管束	radial vascular bundle	放牧草地,	grazing land, pasture,
放射性同位元素	radioactive isotope	放牧地	grazed sward
放射線, 放射,	radiation	放牧地,	grazing land, pasture,
照射		放牧草地	grazed sward
放射線育種	radiation breeding	放牧適性	aptitude for grazing
放射組織	ray	放牧頭数	head of grazing
放出,	emission		animals
エミッション		放牧法	grazing system
膨　潤	swelling	放牧密度	stocking density,
防除, 駆除	control		grazing density
飽食給与	full feeding	放牧料	grazing fee
防除効果	control effect	泡まつ【沫】性	bloat
放任受［授］粉,	open pollination	鼓ちょう【脹】	
開放受［授］		［腸］症,	
粉		鼓ちょう【脹】	
防風垣	wind hedge	［腸］症	
防風林	wind-break forest	ほう【苞】葉,	bract
防腐剤	antiseptic,	ほう【苞】	
	preservative	飽　和	saturation
放　牧	grazing, stocking	飽和脂肪酸	saturated fatty acid
放牧圧	grazing pressure	ホールクロップ	whole crop
放牧育成	rearing on pasture	ホールクロップ	whole-crop silage
放牧家畜	grazing animal	サイレージ	
	［livestock］	母株, 母本	mother plant
放牧カレンダー	grazing calendar	補完関係	complementary
放牧間隔	grazing interval		relation
放牧管理	grazing management	補完草地	complementary
放牧期間,	grazing period		grassland
滞牧日数		牧柵, 柵	fence
放牧技術	grazing technique	牧　場	ranch, livestock farm
放牧季節	grazing season	牧場経営	ranching
放牧牛	grazing cattle	《農業類型》	
放牧強度	grazing intensity	牧場経営者	rancher
放牧権	pasturage right	牧　草	herbage ［pasture］
放牧効果	grazing effect		plant
放牧行動	grazing behavior	牧草乾燥機	hay dryer
放牧最適化仮説	grazing optimization		
	hypothesis		

牧草地	sown [tame, artificial] pasture	補償点	compensation point
撲滅，根絶	eradication	ほ【圃】場容水量	field capacity
牧野《米国の自然草地》	range	補助金	subsidy
		補助剤	adjuvant
牧野組合	pastoral association	補助色素	accessory pigment
牧野樹林	grassland forest	補助飼料	supplementary [supplemental] feed, supplement
牧野林，混牧林	grazing forest [woodland]		
牧養型	grazing type	補助飼料給与，飼料補給	supplemental [supplementary] feeding
牧養力	grazing [carrying, stocking] capacity		
母茎	mother [original, parent] tiller	保水性[力]	water holding capacity
母系（系統）	maternal line	保水力[性]	water holding capacity
母系選抜	maternal line selection	穂数	number of ears [panicles]
保護	protection		
歩行型体重計	walk throw type weigher	歩数記録装置，万歩計	pedometer
補酵素	coenzyme	ホスト，宿主	host
歩行速度	walking speed	補正，調整，調節	adjustment
歩行トラクタ	walking tractor		
保護ケージ	protect cage	保全，保存	conservation
保護作物	nurse crop	捕捉《草の》	prehension
保護樹林《林業》	nurse trees	捕捉作物	catch crop
保護地域	protected area	穂揃い	full heading
母子放牧《牛》	cow/calf grazing	保存，貯蔵	preservation, conservation, storage
母子放牧《緬羊》	ewe/lamb grazing		
母樹	mother tree, seed bearer		
		保存，保全	conservation
母集団，個体群，集団	population	牧区	paddock
		ボディコンディションスコア	body condition score
ほ【圃】場，畑（地）	field		
		保毒[菌]者，担体，キャリア	carrier
ほ【圃】場乾燥	field curing, field drying		
ほ【圃】場試験，野外試験	field experiment [test, trial]	ポドゾル土	Podzol
		ボトムアンローダ	bottom unloader
ほ【圃】場試験法	field plot technique	ボトムプラウ	bottom plow
保証種子	certified seed	ほ【哺】乳	nursing, suckling

穂発芽	sprouting on the panicle, vivipary	埋土種子	buried seed
穂ばらみ	booting	-3/2乗則《自然間引きの》	-3/2 power law of natural thinning
穂ばらみ期	boot stage	まき牛繁殖，自然交配	free breeding on pasture, pasture breeding
穂ばらみ茎	booting tiller [stem]		
ほふく【匍匐】茎	creeper, runner, stolon	播き幅	drill (seeding, sowing) width
ほふく【匍匐】型(の)	prostrate, creeping, stoloniferous	播きみぞ【溝】	seed furrow [stripe]
母平均	population mean	マグネシウム	magnesium, Mg
母本，母株	mother plant	枕　地	headland
穂密度	ear density	マクロシードペレット	macro seed-pellet
ホメオスタシス，生体恒常性	homeostasis	マサ土	Masa, granitic regosol
ホモ[同型]接合体	homozygote	混ぜ播き，混播	mix seeding [sowing]
		マット形成	mat formation
ポリジーン，多遺伝子，微動遺伝子	polygene	マッピング，地図化	mapping
		間引き	thinning
ポリネータ，送粉者，花粉ばい【媒】介者	pollinator	マメ科	Leguminosae
		マメ科の	leguminous
		マメ科(牧)草	herbage [forage] legume
ホルスタイン(種の)	Holstein	マメ科率	ratio of legumes
		マルコフ遷移モデル	Markovian succession model
ま			
		マルチ，被覆	mulch
マーカー	marker	マルチ栽培，被覆栽培	mulch culture, mulching cultivation
マイクロソーム，ミクロゾーム《微粒体》	microsome		
		マルチング，敷草，けい【畦】面被覆	mulching
マイクロプロット	microplot		
		マルトース，麦芽糖	maltose
マイクロボディ，微小体	microbody	マンガン	manganese, Mn
マイコトキシン，かび毒	mycotoxin	マンナン	mannan
		マンノース	mannose
マイコプラズマ様微生物，ファイトプラズマ	phytoplasma	万歩計，歩数記録装置	pedometer

み

見かけの光合成，純光合成	apparent [net] photosynthesis
ミクロゾーム《微粒体》，マイクロソーム	microsome
ミクロフローラ，微生物相	microflora
未経産雌牛，若雌牛	heifer
未耕土	virgin soil
未熟たい【堆】肥	incomplete manure, immature barnyard manure
未熟土	regosol
未熟な，未成熟の	immature
未熟はい【胚】	immature embryo
実生，芽ばえ	seedling
実生選抜	seedling selection
実生［幼植物］の定着	seedling establishment
水管理	water management
水吸収	water uptake
水収支	water budget [balance]
水ストレス	water stress
水ストレス耐性	water stress resistance
水のかん【涵】養，水保全	water conservation
水保全，水のかん【涵】養	water conservation
水ポテンシャル	water potential
水利用	water use
水利用効率	water use efficiency
未成熟の，未熟な	immature
みぞ【溝】	ditch
密植	close [dense] planting
密生（草生）	thick stand
密度，コンシステンシー，粘ちょう【凋】度	consistency
密度	density
密度依存	density dependent
密度効果	density effect
密度独立	density independent
ミトコンドリア	mitochondrion [pl. mitochondria]
ミニマムティレッジ	minimum tillage
ミネラル，無機物	mineral
ミネラル吸収	mineral uptake
脈，静脈	vein
ミルカ，搾乳機	milker

む

向刈り	reaping against lodging
無隔菌糸	aseptate hypha [pl. aseptate hyphae]
無角和牛	Japanese Polled cattle
無機栄養	mineral nutrition
無機栄養生物，独立栄養体，独立栄養生物	autotroph
無気呼吸，嫌気呼吸	anaerobic respiration
無機質肥料	mineral fertilizer
無機成分	inorganic component, mineral
無機物，ミネラル	mineral
麦わら【藁】	wheat straw
無菌化，滅［殺］菌，滅菌法	sterilization

無菌培養	sterile [aseptic] culture	目土	joint soil
無限花序	indefinite [indeterminate] inflorescence	メトヘモグロビン	methemoglobin, MHb
		メトヘモグロビン血液	methemoglobinemia
無効分げつ	non-productive tiller	芽ばえ，実生	seedling
ムコ多糖	mucopolysaccharide	雌花	female [pistillate] flower
無作為化	randomization		
無作為抽出	random sampling	目盛補正，較正，キャリブレーション	calibration
無作為配置	random arrangement		
無脂固形物	solids-not-fat, SNF		
虫こぶ，虫えい	gall	免えき【疫】(性)	immunity
無水ケイ酸，シリカ	silica	綿実かす【粕】	cottonseed meal
		面積	area
無性時代，不完全時代	imperfect state	めん【緬】羊，羊	sheep
無性生殖	asexual reproduction		
無脊椎動物	invertebrate	**も**	
無霜期間	length of frost free season		
		毛管孔げき【隙】	capillary pore
無はい【胚】乳種子	exalbuminous seed	毛管水	capillary water
無病徴感染	symptomless infection	毛根，毛状根	hairy root
無胞子生殖	apospory	毛状根，毛根	hairy root
無融合生殖，アポミクシス	apomixis	網状脈	netted vein
		盲腸	cecum [pl. caecum]
		モーア，刈取機	mower
め		モーアコンディショナ	mower conditioner
芽	bud	木化，リグニン化	lignification
明発芽，光発芽	light germination		
明反応	light reaction	木部	xylem
めしべ，雌ずい	pistil	木部柔組織	xylem parenchyma
雌めん【緬】羊	ewe	木本採食性動物	browser
滅[殺]菌，滅菌法，無菌化	sterilization	木本類	woody [arboreous] plant, arbor
滅菌法，滅[殺]菌，無菌化	sterilization	モグラ	mole
		モグラ暗きょ【渠】せん孔機	mole drainer
メッシュ図	mesh figure [map]		
メッシュデータ	mesh data	モジュール	module
メッセンジャーRNA	messenger RNA, m-RNA	モジュール植物	modular plant

持ち越し	carry-over	[保護]	
もち【糯】性の	glutinous, waxy	野生動物	wild animal
モデリング	modeling	野草，野生植物	wild plant
モデル	model	野草地	native grassland [pasture], rangeland
もどし交雑	back-cross		
もと種，原種《作物》	original [stock] seed	野草放牧地	native pasture, wild grassland used for grazing
モニタリング	monitoring		
モニタリング機器	monitoring equipment		
もみ【籾】	unhulled rice		ゆ
もみ【籾】殻	chaff, hull, husk		
もみ【籾】すり	husking	油圧装置	hydraulic system
盛　土	banking	優位，優占，優性	dominance
モリブデン	molybdenum, Mo		
問題植物	problem plant	有意差	significant difference
		有意水準	level of significance, significant level
	や		
		有意性検定	significance test
野外試験，ほ【圃】場試験	field experiment [test, trial]	誘引剤[物質]	attractant
		有害雑草	noxious weed
夜間放牧	nighttime grazing	有害生物	pest
山　羊	goat	有害廃棄物	hazardous waste
焼　印	brand(ing)	有機栄養生物，従属栄養生物，従属栄養体	heterotroph
焼　畑	burnt field		
焼畑農耕	shifting cultivation		
やく【葯】	anther	有機酸	organic acid
薬　害	chemical (spray) injury	有機質肥料	organic fertilizer
		有機組成	organic composition
薬剤散布［処理］	chemical spraying [application]	有機農業[農法]	organic agriculture [farming]
薬剤防除，化学的防除	chemical control	有機物	organic matter
		有限花序	determinate inflorescence
やく【葯】培養	anther culture		
薬　浴	dipping	有効エネルギー	available energy
野菜，そ【蔬】菜	vegetable	有効温度，実効温度	effective temperature
野生化	escape		
野生種	wild race	有効集団サイズ	effective population size
野生植物，野草	wild plant		
野生生物	wildlife	有効水分	available moisture
野生生物保全	wildlife conservation	有効数字	significant digit

有効積算温度	effective cumulative temperature
有効土層	effective soil depth
有効分げつ	fertile [productive] tiller
有効リン酸	available phosphoric acid
有色体, 色素体, クロマトフォア	chromatophore
有刺鉄線	barbed wire
有糸分裂	mitosis
雄穂《トウモロコシの》	tassel
雄ずい, 雄しべ	stamen
雄穂除去	detasseling
雄ずい先熟	protandry
優性, 優占, 優位	dominance
優性形質	dominant character
優性効果	dominance effect
有性 [完全] 時代《カビ》	perfect state
有性生殖	sexual reproduction
雄性不ねん【稔】(性)	male sterility
融雪促進	acceleration of melting of snow
優占, 優位, 優性	dominance
優占種	dominant species
優占草種	dominant herb [grass]
優占度	abundance, dominance
遊走子	zoospore
遊走子のう	zoosporangium [*pl.* zoosporangia]
有畜農業	stock-holding agriculture
有てい【蹄】動物	ungulate
誘導, 誘発	induction
誘導期, 静止期	lag phase
誘導結合プラズマ [ICP] 発光分光分析	inductively coupled plasma [ICP] emission spectrometry
誘導路	driveway
有毒植物	poisonous plant
有はい【胚】乳種子	albuminous seed
誘発, 誘導	induction
誘発休眠	induced dormancy
遊歩《家畜の》, アイドリング	idling
有胞子細菌	spore-forming bacteria
遊牧	nomadic pastoralism, nomadism
遊牧民	nomads
有用野草	useful wild herb
遊離水, 自由水	free water
輸送	transport

よ

葉位	leaf position on stem
要因, 因子	factor
要因分析法	factorial analysis
葉えき【腋】	leaf axil
葉化	phyllody
溶解《細胞などの》, 溶菌	lysis
溶解性タンパク【蛋白】質	soluble protein, CPs
葉基 [脚]	leaf base
要求 (量)	requirement
溶菌, 溶解《細胞などの》	lysis
葉茎比	leaf-stem ratio
陽光 [陽生] 植物	sun plant
幼根, 小根	radicle

葉コンダクタンス	leaf conductance	養分吸収	nutrient uptake [absorption]
葉　耳	auricle	養分［栄養］ストレス	nutrient stress
葉質の，葉の，葉状の	foliar	葉分析	leaf analysis
葉　序	phyllotaxis	葉　柄	petiole
葉しょう【鞘】	sheath, leaf sheath	葉　脈	leaf vein, nerve, rib
葉状の，葉の，葉質の	foliar	葉脈緑帯	vein banding
幼植物，幼苗	seedling	葉面散布	foliar application, foliar spray
幼植物［幼苗］検定	seedling [nursery] test	葉面診断	foliar diagnosis
葉　身	lamina [*pl.* laminae], leaf blade	葉面積	leaf area
		葉面積指数	LAI [leaf area index]
		葉面積比	LAR [leaf area ratio]
葉身伸長速度	leaf extension rate	羊毛生産	wool production
幼穂形成期	panicle [young ear] formation stage	陽　葉	sun-leaf
		葉緑素，クロロフィル	chlorophyll
幼穂分化期	ear initiation stage, panicle differentiation stage	葉緑体	chloroplast
		葉緑体タンパク【蛋白】質	chloroplastic protein
容水量	moisture [water] capacity	幼令木［樹］	sapling
用水量	irrigation requirement	予　乾	prewilting, wilting
要水量	water requirement	抑圧，抑制	depression
幼生，幼虫	larva [*pl.* larvae]	抑制，抑圧	depression
陽生［陽光］植物	sun plant	抑制遺伝子	inhibiting [repressor] gene, inhibitor
よう【熔】（成）リン（肥）	fused magnesium phosphate	抑制因子	inhibitory factor
		予　測	prediction
容積重	bulk density	予測モデル	prediction model
葉　舌	ligule	預託牛	depositing cattle
ヨウ素	iodine, I	預託方式	depositing system
要素，元素	element	預託牧場	pasture of deposit cattle
要素欠乏	mineral deficiency		
溶脱（作用）	eluviation, leaching	予　防	prevention
幼虫，幼生	larva [*pl.* larvae]	四価染色体	tetravalent [quadrivalent] chromosome
葉　肉	mesophyll		
幼苗，幼植物	seedling	四染色体遺伝	tetrasomic inheritance
幼苗［稚苗］期	seedling stage	四倍体（の）	tetraploid
幼苗［幼植物］検定	seedling [nursery] test	四輪駆動トラクタ	four-wheel drive tractor
葉部割合	percentage of leaves		

ら

RAPD分析	RAPD [random amplified polymorphic DNA] analysis
ライシメータ	lysimeter
ラインインターセプション法	line interception method
ライントランセクト法	line transect method
ラウンドベーラ, ロールベーラ	round baler, roll baler
ラグーン, 沼	lagoon
酪酸	butyric acid
酪酸発酵	butyric fermentation
酪農家	dairy farmer
酪農（経営）	dairy farming, dairying
酪農廃棄物	dairy waste
落葉, せん【剪】葉, 摘葉処理	defoliation
落葉・落枝, 遺体	litter
落葉広葉樹林	deciduous broad-leaved forest
落葉樹	deciduous tree
落葉樹林	deciduous forest
ラジオオートグラフ	radioautograph
ラジオテレメトリ	radiotelemetry
裸子植物	gymnosperm
裸地	bare land
ラッピング	wrapping
ラップサイレージ	wrapped silage
ラテン方格法	latin squares design
ラメット	ramet
乱塊法	randomized block design
卵細胞, 卵子	egg cell
卵子, 卵細胞	egg cell
卵巣のう腫	ovarian cyst
ランダム［不規則, 機会的］分布	random distribution
ランドサット	LANDSAT
ランドスケープ, 景観	landscape
ランナー	runner

り

リーチング, 浸出, 洗脱	leaching
力学的強度	dynamic strength
リグニン	lignin
リグニン化, 木化	lignification
離層	abscission layer
リター層	litter layer
リター分解	litter decomposition, decomposition of litter
履帯トラクタ	crawler tractor
立地	location, site
立地条件	locational condition, geographical and social condition of location
立毛乾燥飼草	cured standing forage
立毛貯蔵	foggage conservation, stand preservation
離乳	weaning
リパーゼ, 脂肪分解酵素	lipase
り【罹】病性の	susceptible
リボフラビン, ビタミンB_2	riboflavin, vitamin B_2
リモートセンシング	remote sensing

硫安, 硫酸アンモニウム	ammonium sulfate
流 域	catchment basin
流去（量），流出（量）	run-off, efflux, outflow
粒 径	kernel depth [size]
流行病	epidemic disease
硫酸アンモニウム, 硫安	ammonium sulfate
流出（量），流去（量）	run-off, efflux, outflow
粒状肥料	granular fertilizer
流通乾草	commercial hay
流通飼料	commercial feed
流動資本	current capital
流入（量）	influx, inflow
流 亡	run-off loss
領域, 縄張り, テリトリ	territory
利用（効）率	efficiency of utilization, use efficiency
両性花	hermaphrodite
量的遺伝	quantitative inheritance
量的遺伝学	quantitative genetics
量的形質	quantitative character
利用頻度	frequency of utilization
両面気孔葉	amphistomatous leaf
両立性, 適合性, 和合性	compatibility
緑 化	revegetation
緑 肥	green manure
緑 変	greening
緑 葉	green leaf
リン	phosphorus, P
臨界域	critical zone
臨界温度, 限界温度	critical temperature
隣家受粉	geitonogamy
林冠, 草冠, キャノピ	canopy
輪換田	paddy field in paddy-upland rotation
輪換畑	upland field in paddy-upland rotation
輪換放牧	rotational grazing, rotation grazing
りん【鱗】茎	bulb
輪 作	crop rotation, rotation
輪作草地, 短年輪作草地	ley
輪作体系	crop rotation system
リン酸吸収係数	phosphate absorption coefficient
リン酸質肥料	phosphate fertilizer
林 床	forest understory
林床植生	ground vegetation, vegetation in forest understory
林地, 樹林地, 森林	woodland, forestland
林内放牧	forest grazing
林分《個々の群落の》	stand

る

ルースバーン, 放飼式牛舎	free barn, loose barn
ルースベーラ	loose baler
ルートマット	root mat
ルーメン, 第一胃, 反すう【芻】胃	rumen
ルーメン運動	rumen exercise
ルーメン液	rumen liquid
ルーメン［第一胃］内細菌	rumen bacteria

る・れ・ろ・わ [87]

ルーメン（内）消化	rumen digestion
ルーメンバイパス	rumen bypass
ルーメン［第一胃］内発酵	rumen fermentation
ルミノロジー	ruminology

れ

冷温帯	cool temperate zone
冷 害	cool summer damage
冷害抵抗性	cool weather resistance
令［齢］構造	age structure
冷処理, 低温処理	chilling
冷 蔵	cold storage
冷涼地帯	cool zone
レーキ	rake
レーザ	laser
レース，菌系	race
れき【礫】	gravel
れき【礫】耕	gravel culture
レクリエーション	recreation
レチノール, ビタミン A_1	retinol, vitamin A_1
列植え	row planting
劣化, 荒廃, 衰退	deterioration, degradation
劣性遺伝子	recessive gene
劣性形質	recessive character
連作, 連続栽培	continuous cropping
連作（障）害	injury by continuous cropping
連 産	continuous calving
連続かんがい【灌漑】	continuous irrigation
連続栽培, 連作	continuous cropping
連続変異	continuous variation
連続放牧	continuous grazing

ろ

ロウ, ワックス	wax
老化, 加令［齢］	ageing, senescence
ろう孔［管］, フィステル	fistula
労働集約度	labor intensity
労働生産性	labor productivity
労働費	labor expense
ローダ	loader
ロータリ耕	rotary tilling [cultivation]
ロータリ耕うん【耘】機	rotary tiller
ロータリモーア	rotary mower
ロールベーラ, ラウンドベーラ	roll baler, round baler
ロールベール	roll bale
ロールベールサイレージ	round bale silage
六炭糖	hexose
ロジスチック曲線	logistic curve
ロジスチック式	logistic formula [equation]
ロゼット, そう【叢】生	rosette
ロゼット植物	rosette plant
ロット	lot
ロトカ・ボルテラ式	Lotka-Volterra equation

わ

わい【矮】化, すくみ, 発育阻止	stunt

わい【矮】性の, い【萎】縮した	dwarf	ワゴン	wagon
		早生（の）	early maturing
		早生品種	early variety
ワイルドフラワー	wild flower	輪だち【轍】, 装輪トラック	wheel track [rut]
若葉《木本植物》	browse	ワックス, ロウ	wax
若雌牛, 未経産雌牛	heifer	わら【藁】	straw
		わら《トウモロコシなどの》	stover
和牛	Japanese Beef cattle		
枠抽出	quadrat sampling	ワラビ型草地	bracken-type grassland
枠法, 方形区法, 区画法	quadrat method	ワラビ中毒	bracken poisoning
和合性, 両立性, 適合性	compatibility	割り当て草量, 供給草量	forage [herbage, pasture] allowance

II 英和の部
English — Japanese

[凡例]
( )：省略可
[ ]：代替可
【 】：非常用漢字
《 》：説明あるいは注記
pl.：複数形

II 英和の部

A

abandoned [uncultivated] land	放棄地
abiotic factor	非生物的環境
abnormal weather	異常気象
abomasum	第四胃
abortion	発育停止，中絶
aboveground [top, aerial] part	地上部
abscisic acid	アブシジン酸
abscission layer	離層
absorbing [sucking] root	吸収根
absorption, uptake	吸収
abundance, dominance	優占度
abundant [good] harvest	豊作，上作
acaricide	殺だに剤
accelerated generation advancement	世代促進
acceleration of melting of snow	融雪促進
accessory pigment	補助色素
accidental species	偶存種
acclimation, acclimatization, domestication	順化，じゅん【馴】化
acclimation, accustoming, preconditioning	じゅん【馴】致
acclimatization, acclimation, domestication	順化，じゅん【馴】化
accomodation	順応
accumulated [cumulative] air temperature	積算気温
accumulated precipitation	積算降水量
accumulated [cumulative] temperarure	積算温度
accumulation	集積，たい【堆】積，蓄積
accustoming, acclimation, preconditioning	じゅん【馴】致
acervulus [pl. acervuli]	分生子層
acetic acid	酢酸
acetylene reduction assay	アセチレン還元法
acid	酸
acid detergent fiber, ADF	酸性デタージェント繊維
acid detergent insoluble nitrogen, ADIN	酸性デタージェント不溶性窒素
acid manure, acidic fertilizer	酸性肥料
acid plant	酸性植物
acid rain	酸性雨
acid soil	酸性土壌
acid tolerance	耐酸性
acidic fertilizer, acid manure	酸性肥料
acidification	酸性化
acid-insoluble ash	酸不溶性灰分
acidity	酸性，酸度，酸性度
acidosis	アシドーシス
acquired character	獲得形質
action spectrum	作用スペクトル
activation	活性化，ふ【賦】活（化）

activator	ふ【賦】活剤,活性化物質,活性化因子	adventitious root	不定根
		aeration	通気,ばっ【曝】気
active absorption	積極的吸収,能動的吸収	aeration pressure	通気圧
		aerial application	空中散布
active sludge	活性汚泥	aerial [top, aboveground] part	地上部
active substance	活性物質		
active tillering stage	分げつ盛期	aerial root	気根
active transport	能動輸送	aerial seeding [sowing]	航空播種
actual condition survey, investigation of actual condition	実態調査	aerial tiller(ing)	空中分げつ
		aerobe	好気性生物
actual use	実質利用	aerobic bacteria	好気性(細)菌
actual volume	実容積	aerobic deterioration	好気的変敗
acute	急性の	affection, infection	感染
adaptability	適応性	affinity, compatibility	親和性
adaptability to heavy dressing of fertilizer	耐肥性	afforestation	人工造林,植林
		afforested land, tree plantation	植林地,造林地
adaptation	適応		
adaptive response	適応(的)反応[応答]	aftermath	再生草
		aftermath grazing	2番草放牧
addition	添加	after-ripening	後熟
additional [additive] experiment	追加実験	age	年令[齢]
		age in month	月令[齢]
additional manure, topdressing	追肥	age structure	令[齢]構造
		aged [old] grassland, ageing pasture [meadow]	経年草地
additive	添加剤[物]		
additive effect	相加的効果,相加作用	ageing, senescence	加令[齢], 老化
		ageing, maturity, maturation	熟成
additive [additional] experiment	追加実験		
		ageing pasture [meadow], aged [old] grassland	経年草地
adjustment	調整,調節,補正		
adjuvant	補助剤	aggregated distribution	集中分布
adlibitum feeding	不断給餌	aggregation	集合,凝集
adsorption	吸着	aggregation	団粒化
adult, imago	成(個)体,成虫	aggressive	進攻的《生態》
		agricultural chemical, pesticide	農薬
advanced generation	後期世代		
advanced synthetic generation	後期合成世代	agricultural extention worker [officer]	農業改良普及員
adventitious bud	不定芽		

agricultural income	農業所得
agricultural machinery, farm machinery	農業機械
agricultural meteorology	農業気象学
agricultural production	農業生産
agricultural structure	農業構造
agricultural wastes	農業廃棄物
agrobacterium	アグロバクテリウム
agroforestry	アグロフォレストリ
agronomic [economic] character [trait]	農業 [実用, 経済] 形質
agronomy	農学
air conditioning room, climatic chamber	人工気象室
air drainage	排気
air permeability	通気性
air pollution	大気汚染
air temperature	気温
air-borne disease	空気伝染病
air-dry matter	風乾物
air-dry soil	風乾土
air-dry weight	風乾重
airplane spraying	航空機散布
airtight silo	気密サイロ
albedo	日射反射率
albino	アルビノ, 白子, 白色種
albuminous seed	有はい【胚】乳種子
alcohol	アルコール
aleurone layer	アリューロン層, こ【糊】粉層
alfalfa meal	アルファルファミール
alien [invading] plant	侵入植物
alimentary canal, digestive tract	消化管
alkali	アルカリ
alkaline soil	アルカリ土壌
alkali-treated straw	アルカリ処理わら【藁】
alkaloid	アルカロイド
allele	対立遺伝子, アレル, アレレ
alleropathy	他感作用, アレロパシ
alley [stripe] cropping	帯状栽培
alliance	群団
allied [related] species	近縁種
allocation, partition	分配, 配分
allogamous	他殖性 (の)
allogamous plant	他殖性植物
allogamy	他殖, 他家生殖
allogamy, outcrossing, outbreeding	他殖性, 他配
allogamy, cross pollination	他家受粉
allometry, relative growth	相対成 [生] 長
alloploid, allopolyploid	異質倍数体
allopolyploid, alloploid	異質倍数体
all-season silage feeding	通年サイレージ給与
alluvial fan	扇状地, 沖積扇状地
alluvial soil	沖積土
alpine	高山の
alpine grassland	高山草地
alpine [mountain] grazing	山地放牧
alpine meadow	高山採草地
alpine pasture	山地放牧地
alpine region	高山地方
alpine zone	高山帯
alternate	互生《葉》
alternate [intermediate] host	中間宿主

alternate row seeding [sowing]	交互条播	anaerobic respiration	無気呼吸, 嫌気呼吸
alternating temperature	変温	analysis of covariance	共分散分析
alternation of generations, heterogenesis	世代交代	analysis of farm operation	経営分析
alternative agriculture [farming]	代替農業	analysis of variance, ANOVA	分散分析
altitude, elevation	標高	analysis of water stable aggregate	団粒分析
aluminum, Al	アルミニウム	anatomical	解剖学的
amenity	快適性, アメニティ	anatomy	解剖学
		andesite	安山岩
amide	アマイド, アミド	Andosol	黒ボク土
		anemia	貧血(症)
amino acid	アミノ酸	anemometer	風速計
ammonia	アンモニア	anemophily	風ばい【媒】
ammonia plant	アンモニア植物	aneuploid, heteroploid	異数体
ammoniated rice hull [husk]	アンモニア処理もみ【籾】殻	aneuploidy, heteroploidy	異数性
ammonification	アンモニア化成作用	angiosperm	被子植物
		animal [livestock] industry	畜産業
ammonium nitrate	硝酸アンモニウム	animal ecology	動物生態学
ammonium nitrogen	アンモニア態窒素	animal feeding	家畜飼養
		animal husbandry, livestock farming	畜産, 畜産経営
ammonium sulfate	硫酸アンモニウム, 硫安	animal management	家畜管理
amount of excreta	排泄量	animal manure, livestock waste	家畜ふん【糞】尿
amount of growth	生育量		
(amount of) precipitation	降水(量)	animal physiology	家畜生理
		animal product	畜産物
(amount of) solar radiation	日射(量)	animal production	家畜生産
		animal [zootechnical] science, zootechnology	畜産学
amphidiploid	複二倍体		
amphistomatous leaf	両面気孔葉		
amylase	アミラーゼ	animal unit	家畜単位
anaerobic bacteria [microbe]	嫌気性(細)菌	animal welfare	動物福祉
		anion	アニオン
anaerobic condition	嫌気的条件	annual	一年生(の), 当年生
anaerobic degradation	嫌気的分解		
anaerobic microbe [bacteria]	嫌気性(細)菌	annual ring	年輪

A [93]

annual [current-year] tiller [stem]	当年茎	ferlilizer, fertilizer rate	
antagonism	きっ【拮】抗作用, きっ【拮】抗	applied ecology	応用生態学
		aptitude for grazing	放牧適性
		aquatic biomass	水生バイオマス
anther	やく【葯】	aquatic [hydrophytic] weed	水生雑草
anther culture	やく【葯】培養		
antheridium [pl. antheridia]	蔵精器	arable [cultivated] land	既耕地, 耕地
anthesis, blooming, flowering	開 花	arable soil	耕 土
		arbor, woody [arboreous] plant	木本類
anthesis	開やく【葯】		
anthocyan	アントシアン	arboreous [woody] plant, arbor	木本類
antibiotic	抗生物質, 抗菌性の		
		arbuscular mycorrhizae	菌根菌
antibody	抗 体		
antigen	抗 原	arctic zone	寒 帯
antimold	防かび剤	area	面 積
antioxidant	抗酸化剤, 抗酸化の	area covered by dung, dung-deposited place [area]	排ふん【糞】地
antiseptic, preservative	防腐剤		
		arid	乾燥した《土地》
antiserum [pl. antisera]	抗血清		
		arid climate	乾燥気候
apex	頂部, 頂端, せん【尖】部	arid grassland	乾燥草地
		arid land	乾燥地
apical [terminal] bud	頂 芽	arid region [zone]	乾燥地帯
apical dominance	頂芽優勢	aridity	干[旱]ばつ【魃】度
apical meristem	頂端分裂組織		
apomixis	アポミクシス, 無融合生殖	aridity index	干[旱]ばつ【魃】指数
		arista, awn	のぎ, ぼう【芒】
apospory	無胞子生殖	aroma component	香気成分
apparent [net] photosynthesis	見かけの光合成, 純光合成	arthropod fauna	節足動物相
		artificial community, cultivated vegetation	人工群落
apparent specific gravity, bulk density	仮比重		
		artificial crossing	人工交雑, 人工交配
appearance, emergence	出 現		
		artificial [in vitro] digestibility	人工消化率
appetite	食 欲		
application	施 用	artificial drying	人工乾燥
application rate of	施肥量		

artificial [sown] grassland	人工草地	auricle	葉耳
artificial impact [pressure]	人為圧	autoecology, ideoecology	個生態学
		autogamous plant	自殖性植物
artificial inoculation	人工接種	autogamy, self-fertilization, selfing	自家受精，自殖
artificial insemination, AI	人工授精		
artificial lighting	人工照明	autohexaploid	同質六倍体
artificial mutation	人工突然変異，人為的突然変異	Automated Meteorological Data Acquisition System, AMeDAS	アメダス
artificial [sown, tame] pasture	牧草地		
		automatic [self] feeder	自動給餌器
artificial pollination	人工授粉	automatic control	自動制御
artificial pressure [impact]	人為圧	automatic measurement	自動計測
artificial seed	人工種子	automatic waterer	自動給水器
artificial suckling	人工哺乳	autoploid, autopolyploid	同質倍数体
ascorbic acid, vitamin C	アスコルビン酸，ビタミンC	autopolyploid, autoploid	同質倍数体
aseptate hypha [*pl.* aseptate hyphae]	無隔菌糸	autoradiograph	オートラジオグラフ
		autotetraploid	同質四倍体
aseptic [sterile] culture	無菌培養	autotroph	無機栄養生物，独立栄養体，独立栄養生物
asexual reproduction	無性生殖		
ash	灰分		
assessment, evaluation	評価	autumn [fall] crop	秋作
assimilation	同化（作用）	autumn manuring, fall dressing	秋肥
assimilation chamber method [technique]	同化箱法		
		autumn saved pasture, ASP	備蓄用牧区
assimilation product	同化産物		
assimilatory organ	同化器官	autumn [fall] seeding [sowing]	秋播き
assimilatory starch	同化デンプン【澱粉】		
		auxin	オーキシン
association, community	群落，群集	available energy	有効エネルギー
		available moisture	有効水分
atmospheric model	大気モデル	available phosphoric acid	有効リン酸
atomic absorption spectrochemical analysis	原子吸光法		
		avirulence	非病原性，弱毒性
attractant	誘引剤［物質］	awn, arista	のぎ，ぼう【芒】

axial [tap] root	直根
axillary bud	腋芽

B

β-carotene	β-カロチン
back-cross	もどし交雑
bacterial disease	細菌病
bacterial exudation	細菌漏出
bacterial ooze	細菌粘塊
bacterium [pl. bacteria]	細菌
bag silo	バッグサイロ
bagasse	バガス《さとうきびの搾りかす【粕】》
balance, budget	収支
balance	出納
bale	こん【梱】包, ベール
bale loader	ベールローダ
bale thrower	ベールスローワ
bale wagon	ベールワゴン
baled silage	こん【梱】包サイレージ
baled hay	こん【梱】包乾草
band plowing [tilling]	帯状耕うん【耘】
band seeding [sowing]	帯状播種
band tilling [plowing]	帯状耕うん【耘】
banking	盛土
barbed wire	有刺鉄線
bare land	裸地
barium, Ba	バリウム
bark	樹皮
bark compost	バークたい【堆】肥, 樹皮たい【堆】肥
barn, stable	畜舎
barn	農具舎
barn cleaner	バーンクリーナ
barnyard [stable] manure, manure	畜肥, きゅう【厩】肥, 肥［こえ］
barrier	障害物
basal application of fertilizer, basal dressing	基肥
basal area	基底面積
basal cover [coverage]	基底被度
basal diet [feed]	基礎飼料
basal dressing, basal application of fertilizer	基肥
basal metabolism	基礎代謝
basalt	玄武岩
base	塩基
base	基部
base [cation] exchange capacity, CEC	塩基交換容量
base saturation percentage	塩基飽和度
basic amino acid	塩基性アミノ酸
basic fertilizer	塩基性肥料
basic vegetative growth	基礎栄養成［生］長
basidiomycetes	担子菌（類）
basidiospore	担子胞子
basidium [pl. basidia]	担子柄
bedding, litter	敷料, 敷わら【藁】
beef cattle	肉（用）牛
beef cattle farming	肉用牛経営
beef production	肉牛生産［畜産］
beet pulp	ビートパルプ
behavior	行動, 挙動
behavior pattern	行動型
behavioral ecology	行動生態学

belowground, [subterranean, underground] part	地下部	biometrics	生物測定学
		bioreactor	バイオリアクタ
		bioremediation	生物環境修復, バイオリメディエーション
biennial	二年生（の）		
big roll baler, big round baler	ビッグロールベーラ, ビッグラウンドベーラ		
		biosphere	生物圏, バイオスフェア
big round baler, big roll baler	ビッグラウンドベーラ, ビッグロールベーラ	biosynthesis	生合成
		biota	生物相
		biotechnology	バイオテクノロジ
bile	胆汁	biotelemetry	バイオテレメトリ
binder	刈取結束機		
binder	結合剤	biotic community	生物共同体
binding species	共存種	biotic factor	生物要因
binomi(n)al distribution	二項分布	biotic indicator	生物指標
		biotope	ビオトープ
bioassay	生物（学的）検定, バイオアッセイ	biotron	バイオトロン
		biotype, ecotype	生態型, エコタイプ
biochemical oxygen demand, BOD	生物化学的酸素要求量	bird injury	鳥害
		birth rate	出生率
biochemistry	生物化学, 生化学	bite	食いちぎり, か【噛】み, バイト
bioclimate	生物気候		
biodiversity	生物多様性	bite area	バイト面積
bioecology	生物生態学	bite depth	バイト深
bioeconomics	生物経済学	bite mass	バイト量
biological control	生物（学）的防除［除去］	bite [biting] rate	バイト速度
		bite size	バイトサイズ
biological spectrum	生活型組成, 生活形スペクトル	bite volume	バイト容積
		bite weight	バイト重
		biting [bite] rate	バイト速度
biological value, BV	生物価	bivalent chromosome	二価染色体
biomass	生物（体）量, バイオマス	blended [mixed] fertilizer, composed manure	配合肥料
biomass, standing crop	現存量		
biomass production	バイオマス生産	Blix	ブリックス糖度
biome	バイオーム《生物群系》		
biometer	生物計		

bloat	鼓脹［腸］症, 泡まつ【沫】性鼓脹［腸］症	bractlet	小苞
		branch(ing)	分枝
		branching form [type]	分枝型
		branching type [form]	分枝型
blood	血液	brand(ing)	焼印
blood glucose	血糖	brans	ぬか類
blood-sucking insect	吸血昆虫	breaker [breaking] plow	新墾プラウ
blooming, anthesis, flowering	開花	breaking [breaker] plow	新墾プラウ
blooming stage, flowering time	開花期	breeder('s) seed	育種家種子
blower	送風機, ブロワ	breeding	育種, 繁殖《動物》
body condition score	ボディコンディションスコア	breeding, reproduction	繁殖《動物》, 生殖《動物》
body temperature	体温		
body [live] weight gain	増体, 体重増加	breeding [reproductive] cattle	繁殖牛
bog, moor	湿原	breeding season	繁殖季節
bolting	抽たい【苔】	breeding stock	種畜
boot [flag] leaf	止葉	brewer's dried yeast	乾燥ビール酵母
boot stage	穂ばらみ期	brewer's grain	ビールかす【粕】
booting	穂ばらみ		
booting tiller [stem]	穂ばらみ茎	broad ridge	広うね【畦】
border [edge] effect	周縁効果	broadcast(ing)	散播
boron, B	ホウ素	broadcaster	散播機
botanical [floristic] composition	種組成《植物》, 植物構成, 草種構成, 草種割合	broadcasting of fertilizer	表面散布
		broad-leaved forest	広葉樹林
		broad-leaved tree	広葉樹
botany	植物学	broad-leaved weed	広葉雑草
bottom grass, undergrowth, understory	下草, 下繁草, 下層植生	brown forest soil	褐色森林土
		browning	褐変
		browse	若葉《木本植物》
bottom grazing	下層採食		
bottom plow	ボトムプラウ	browse [defoliation] tolerance	被食耐性
bottom unloader	ボトムアンローダ		
		browser	木本採食性動物
bovine	牛（の）	browsing	新芽（をくう）
bracken poisoning	ワラビ中毒	brush [bush] control	かん【灌】木除去
bracken-type grassland	ワラビ型草地		
bract	ほう【苞】, ほう【苞】葉	buck [sweep] rake	スィープレーキ
		bucket dozer	バケットドーザ

bucket elevator	バケットエレベータ	organization	
bud	芽	butter fat [milk fat] percentage	乳脂率
budding, emergence	出芽, ほう【萌】芽	butyric acid	酪酸
		butyric fermentation	酪酸発酵
budding	出らい【蕾】	by-pass protein	バイパスタンパク【蛋白】質
budding time	出芽期		
budget, balance	収支		
buffalo	バッファロ, 水牛	by-product	副産物

C

buffer	バッファ		
buffer action	緩衝作用		
buffering capacity	緩衝能	C-F ratio	C-F比
bulb	りん【鱗】茎	C_3 pathway	C_3 経路
bulk [mass] emasculation	集団除雄	C_3 plant	C_3 植物
		C_4 dicarboxylic acid cycle	C_4-ジカルボン酸回路
bulk density, apparent specific gravity	仮比重	C_4 pathway	C_4 経路
		C_4 plant	C_4 植物
bulk density	容積重	cadmium, Cd	カドミウム
bulky feed, roughage	粗飼料	cafeteria feeding, free choice feeding	自由選択給餌
bull	雄牛, 種牛		
bumper gate	木戸《放牧場の》	cafeteria method	キャフェテリア法, 自由選択法
bunch, tufted, tussock	株型の		
bunch [tufted] type	そう【叢】状型, そう【叢】生型, 株型	cage	ケージ
		calcareous soil	石灰質土壌
		calcicole, calciphilous plant	石灰植物
bunker silo	バンカーサイロ		
buried seed	埋土種子	calciferol, vitamin D	カルシフェロール, ビタミンD
burned-over land	火入れした土地		
burning	火入れ		
burnt field	焼畑	calcification	石灰集積作用, カルシウム沈着
bush cutter	刈払い機, ブッシュカッタ		
		calcined phosphate	焼成りん肥
bush harrow	柴ハロー	calciphilous plant, calcicole	石灰植物
business [farm] accounting	経営計算		
		calciphobous plant	嫌石灰植物
business [management] balance	経営収支	calcium carbonate	炭酸カルシウム
		calcium cyanamide	石灰窒素
business [farm]	経営組織		

calcium fertilizer, liming material	石灰質肥料 [資材]	carbon-nitrogen ratio, C-N ratio	C-N比, 炭素窒素比
calcium, Ca	カルシウム	care, management	管理
calf, veal calf	子牛	carnivorous insect	肉食性昆虫
calf starter	子牛用離乳飼料	carotenoid	カロチノイド
calibration	目盛補正, 較正, キャリブレーション	carrier	担体, キャリア, 保毒[菌]者
calibration equation	検量線	carrying [grazing, stocking] capacity	牧養力
callus	カルス		
callus induction	カルス誘導	carry-over	持ち越し
Calvin cycle	カルビン回路	caryopsis	えい【穎】果,
calving, parturition	分娩		えい【穎】花
calyx, vase	がく	castration, emasculation	除雄《植物》, 去勢《動物》, 性腺摘除《動物》
cambium	形成層		
Cambridge roller	ケンブリッジローラ		
canopy	草冠, 林冠, キャノピ	catch crop	捕捉作物
		catchment area	集水面積
canopy, crown	樹冠	catchment basin	流域
canopy [crown] closure	うっ閉《林業》	cation	カチオン
		cation [base] exchange capacity, CEC	塩基交換容量
canopy structure	草冠構造		
capacity	受容力	cattle	牛
capillary pore	毛管孔げき【隙】	cattle clothes	牛衣
		cattle fattening farming	肉用牛肥育経営
capillary water	毛管水		
capsule	カプセル	cattle feces	牛ふん【糞】
carbohydrate	炭水化物	cattle manure	牛ふん【糞】尿, 牛ふん【糞】たい【堆】肥
carbohydrate metabolism	炭水化物代謝		
carbon, C	炭素	cattle slurry	牛スラリー
carbon dioxide [CO_2]	炭酸ガス, 二酸化炭素	cattle trail [track]	牛道
		cattle urine	牛尿
carbon dioxide [CO_2] assimilation	炭酸同化	causal agent, pathogen	病因, 病原(体)
		causal organism	病原体
carbon dioxide [CO_2] fixation	炭酸固定	cecum [*pl.* caecum]	盲腸
		ceiling leaf area index	最大葉面積指数
carbon isotope discrimination	炭素同位体分別	cell	細胞
		cell culture	細胞培養
carbon sink	炭素シンク	cell differentiation	細胞分化
carbon source	炭素源	cell division	細胞分裂

cell membrane	細胞膜
cell wall	細胞壁
cell wall component [constituent], CWC	細胞壁成分
cell wall substances, CW	細胞壁物質
cell(ular) contents, CC	細胞内容（物）
cellulase	セルラーゼ
cellulose	セルロース
cereal	穀物，か【禾】穀類
cereal grains	穀類
certified seed	保証種子
chaff, hull, husk	もみ【籾】殻
chain conveyer	チェーンコンベア
chain harrow	チェーンハロー
chamaephytes	地表植物
change-over design	反転法
changes	変遷
channel	水路，チャンネル
character	形質，特性
characteristic	特性（の）
characteristic species	標徴種
check variety	対照品種，比較品種
cheese [shade] cloth	寒冷紗
chelate	キレート
chemical (spray) injury	薬害
chemical additive	化学添加剤
chemical component	化学成分
chemical composition	化学組成
chemical control	化学的防除，薬剤防除
chemical fertilizer	化学肥料
chemical oxygen demand, COD	化学的酸素要求量
chemical spraying [application]	薬剤散布［処理］
chemical treatment	化学処理
chewing, mastication	そしゃく【咀嚼】
chewing time	そしゃく【咀嚼】時間
chi square test	カイ自乗検定
chilling	冷処理，低温処理
chilling requirement	低温要求度
chimera	キメラ
chisel plow	チゼルプラウ
chitin	キチン
chlamydospore	厚膜胞子，厚壁胞子
chlorine, Cl	塩素
chlorophyll	葉緑素，クロロフィル
chloroplast	葉緑体
chloroplastic protein	葉緑体タンパク【蛋白】質
chlorosis	黄変，白化，クロロシス
cholesterol	コレステロール
chop length	切断長
chopper	細切機
chopping	細切
chromatid	染色分体
chromatin	染色質
chromatophore	有色体，色素体，クロマトフォア
chromogen	クロモーゲン，色素原
chromoprotein	色素タンパク【蛋白】質
chromosomal mutation	染色体突然変異
chromosome	染色体，クロモソーム
chromosome map	染色体地図
chromosome number	染色体数

circulation of material, material [matter] cycle	物質循環
clamp silo	クランプサイロ
classification	分類
clay	しょく【埴】土, 粘土
clayey loam	しょく【埴】壌土
clear cutting	皆伐《林業》
cleistogamy	閉鎖花受精
climate change	気候変動, 気候変化
climatic [meteorological] element	気候要素
climatic chamber, air conditioning room	人工気象室
climatic factor	気候要因
climatic index	気候指数
climatic province	気候区
climatic zone	気候帯
climax	極相, 極期, クライマックス
climax community [vegetation]	極相群落, クライマックス群集［植生］
climax dominants	極相群落優占種, クライマックス優占種
climax forest	極相林
climax vegetation [community]	極相群落, クライマックス群集［植生］
climograph	気候図
clipping method	切えい【穎】法
clod	土塊
clod breaking, soil crushing	砕土
clod crusher, disk harrow	砕土機, ディスクハロー
clonal growth	クローン成［生］長, クローナル成［生］長
clonal propagation	栄養系繁殖, クローン繁殖
clonal selection	栄養系選抜
clonal separation	栄養系分離
clone	栄養系, クローン
clone structure	クローン構造
cloning	クローン化
close [dense] planting	密植
closed pasture	禁牧草地
closing the grazing, end of grazing, housing from pasture	退牧
club root	根こぶ
clumped germination	集中的発芽
cluster analysis	クラスター分析
C-N ratio, carbon-nitrogen ratio	C-N比, 炭素窒素比
CO_2 elevation, elevated CO_2	CO_2 上昇
CO_2 enrichment	CO_2 施肥
coaction	生物相互作用
coast vegetation	海岸植生
coated [coating] seed	コート種子
coating	被覆, コーティング
coating [coated] seed	コート種子
cob, rachis	穂軸
cobalt, Co	コバルト
coefficient of correlation, correlation coefficient	相関係数
coefficient of homogeneity, CH	均質度係数
coefficient of variation, CV	変異［変動］係数
coenzyme	補酵素

coevolution	共進化
coexistence, cooccurrence	共存
colchicine	コルヒチン
cold area [region, district]	寒冷地
cold damage [injury]	寒害
cold hardiness, cold resistance [tolerance]	耐寒性, 耐冷性
cold resistance [tolerance], cold hardiness	耐寒性, 耐冷性
cold storage	冷蔵
coleoptile	子葉しょう【鞘】, しょう【鞘】葉
coleorhiza	根しょう【鞘】
coliform bacilli	大腸菌
collection	採集
colonization	すみつき, 侵入・定着
colonizing ability	すみつき能力, 侵入・定着能力
colony	菌そう【叢】, コロニー, 群体
color tone	色調
colorimetric analysis	比色分析
colostrum	初乳
colter harrow	なたハロー
column chromatography	カラムクロマトグラフィ
combining ability	組合せ能力
commercial feed	流通飼料
commercial hay	流通乾草
commercial variety	市販品種, 実用品種
common [communal] land	共有地, 入会地
common [communal] pasture	共同放牧地
common plow	単用プラウ
communal [community] forest	公有林
communal [common] land	共有地, 入会地
communal [common] pasture	共同放牧地
community	共同体, 共同社会
community, association	群落, 群集
community dynamics	群集動態（学）
community ecology	群集生態学
community [communal] forest	公有林
community measures	群落測度
community structure	群落構造
compacting, firming, soil packing	鎮圧
compacting, trampling, treading	てい【蹄】傷, 踏圧, 踏付け
compaction, consolidation	圧密, 締固め
compactness	ち【緻】密度
companion cell	伴細胞
companion crop	同伴作物
companion grass	同伴イネ科牧草
companion [mixed] planting	混植
companion species	同伴種
compatibility	両立性, 適合性, 和合性
compatibility, affinity	親和性
compensation point	補償点
compensatory growth	代償性発育
competition	競合［争］, 競合い

competition of grass, interspecific competition of grass	草種競合	conservation, preservation, storage	貯蔵，保存
competition-density effect, C-D effect	競争密度効果	conservation	保全，保存
competitive ability	競争力	consistency	コンシステンシー，粘ちょう【凋】度，密度
competitive exclusion	競争排除		
complementary grassland	補完草地		
complementary relation	補完関係	consolidation, compaction	圧密，締固め
complete dominance	完全優勢	consolidation of land	土地交換分合
composed manure, blended [mixed] fertilizer	配合肥料	constance	恒存度，常在度
		constant species	恒存種
		constant [fixed] temperature	定温，恒温
composite sample	混合試料	constituent, composition	組　成
composition, constituent	組　成	consumer	消費者
composition of feed, feed composition	飼料成分組成	contaminated soil	汚染土壌
		content	含有量，含量，含有率
compost	コンポスト		
compost and barnyard manure	たいきゅう【堆厩】肥	continuous calving	連　産
		continuous cropping	連作，連続栽培
composting	たい【堆】肥化，コンポスト化，発酵乾燥	continuous grazing	連続放牧
		continuous irrigation	連続かんがい【灌漑】
compound fertilizer	化成肥料	continuous snow cover duration	根雪期間
compound leaf	複　葉		
compressed [processed] hay	成形乾草	continuous variation	連続変異
		contour cropping	等高線栽培
concentrate (feed)	濃厚飼料	contour strip cropping	等高帯栽培
concentration	濃度《溶液の》	control	防除，駆除
concentration	濃縮（物）	control effect	防除効果
conception rate	受胎率	control plot	対照区
condition diagnosis	状態診断	controlled [restricted] feeding	制限給餌
conduction	伝導，導通		
conductive tissue	通導組織	controlled experiment	制御実験 [試験]
confidence interval	信頼区間		
conidiophore	分生子柄	controlled release fertilizer	肥効調節型肥料
conidium [pl. conidia]	分生子		
conifer	針葉樹	convection	対　流
coniferous forest	針葉樹林		

conventional grazing (system)	慣行放牧	correlation analysis	相関分析
convergence	収れん【斂】, 収束, 集合	correlation coefficient, coefficient of correlation	相関係数
conversion efficiency	変換効率	cortex	皮層, 皮質
converted paddy field	水田転換畑	cost	コスト
cooccurrence, coexistence	共　存	cost of farming, expenditure for agricultural management, farm management cost	農業経営費
cool season [temperate] grass	寒地型イネ科牧草		
cool summer damage	冷　害		
cool temperate zone	冷温帯		
cool weather resistance	冷害抵抗性	cottonseed meal	綿実かす【粕】
		cotyledon	子　葉
cool zone	冷涼地帯	coumarin	クマリン
cooperation	協同, 協同作業	covariance	共分散
cooperative [public] grassland	公共草地	cover crop	被覆作物
		cover degree	被覆度
cooperative [public] pasture, cooperative [public] stock farm	公共牧場	coverage	被　度
		covering with soil, molding	覆　土
cooperative [public] stock farm, cooperative [public] pasture	公共牧場	cow	成雌牛
		cow/calf grazing	母子放牧《牛》
		cow-calf farming	肉用牛繁殖経営
		cow-calf finishing operation	繁殖肥育一貫経営
copper, Cu	銅		
coppice [sprout] forest	ほう【萌】芽林《林業》	cow-day, CD	カウデー
		crawler tractor	履帯トラクタ
cork formation	コルク形成	creep feed	餌づけ飼料
corm	球茎	creep grazing	クリープ放牧
corn harvester	トウモロコシ収穫機, コーンハーベスタ	creeper, runner, stolon	ほふく【匍匐】茎
		creeping, prostrate, stoloniferous	ほふく【匍匐】型(の)
corn planter	トウモロコシ点ま【蒔】き機, コーンプランタ	critical day length	限界日長
		critical temperature	限界温度, 臨界温度
		critical yield	収量限界
corn stover	トウモロコシかん【稈】	critical zone	臨界域
		crop	作　物
corolla tube	花管	crop growth rate, CGR	個体群成長速度[生]
correlation	相関		

crop residue	作物残渣, 作物遺体	crude undecomposed organic matter	粗大有機物
crop rotation, rotation	輪作	crumbled [aggregate] structure	団粒構造
crop rotation system	輪作体系		
crop situation	作柄	crush	圧搾, 粉砕
cropping order [sequence]	作付順序	crush	破砕
		crusher	圧搾機
cropping ratio	作付比率	crusher	圧ぺん機
cropping season	作期	crushing mill	粉砕機
cropping [cultivation] system	作付方式, 栽培[作付]体系	crust	土膜
		crust	外被, か皮, 殻皮
cross [hybridization] breeding	交雑育種	cube	キューブ
		culm, stalk	かん【稈】
cross combination	交配組合せ	culm length	かん【稈】長
cross fertilization	他家受精	culti-packer	カルチパッカ
cross incompatibility	交雑不和合性, 交配不和合性	cultivar	栽培品種, 品種
		cultivated [growing] field	栽培ほ【圃】場
cross pollinating crop	他殖性作物, 他家受粉作物	cultivated [arable] land	既耕地, 耕地
cross pollination, allogamy	他家受粉	cultivated species	栽培種
		cultivated vegetation, artificial community	人工群落
cross protection	交差耐性, 干渉作用		
		cultivation, culture	栽培, 耕作
crossbred, hybrid	交雑種, 雑種	cultivation, plant husbandry	栽培管理
crossing, hybridization	交雑, 雑種形成		
crossing over	交叉, 乗換え, 交差	cultivation, plowing and harrowing, tillage	耕うん【耘】
crossing rate	交雑ねん【稔】性, 交雑率		
		cultivation condition	栽培条件
crown	冠部	cultivation layer	作土
crown, canopy	樹冠	cultivation limit	栽培限界
crown [canopy] closure	うっ閉《林業》	cultivation [plant husbandry] method	栽培法
crown [nodal] root	冠根		
crown cover	冠部被度	cultivation [cropping] system	作付方式, 栽培[作付]体系
crown density	うつ閉度		
crude ash	粗灰分		
crude fat	粗脂肪	cultivation system [pattern]	栽培様式
crude fiber	粗繊維		
crude protein	粗タンパク【蛋白】質	cultivation technique	栽培技術
		cultivator	カルチベータ

cultural control	耕種的防除	cytology	細胞学
culture, cultivation	栽培，耕作	cytoplasm	細胞質
culture	培養	cytoplasmic genome	細胞質ゲノム
culture medium	培地，培養液	cytoplasmic inheritance	細胞質遺伝
cumulative [accumulated] air temperature	積算気温	cytoplasmic male sterility	細胞質雄性不ねん【稔】
cumulative [accumulated] temperature	積算温度		

D

cured standing forage	立毛乾燥飼草	daily behavior	日周行動
curing	日乾	daily energy requirement	日エネルギー要求量
current capital	流動資本	daily gain	日増体量
current-year [annual] tiller [stem]	当年茎	dairy [milk] performance	泌乳能力
cuticle	クチクラ	dairy breed	乳用種
cuticular layer	クチクラ層	dairy cattle	乳（用）牛
cuticular penetration	角皮貫入	dairy [milking, lactating] cow	搾乳牛，泌乳牛
cutter bar	カッタバー	dairy farmer	酪農家
cutter blower	カッタブロア	dairy farming, dairying	酪農（経営）
cutting	刈取り	dairy waste	酪農廃棄物
cutting	切断	dairying, dairy farming	酪農（経営）
cutting, mechanical (forage) harvesting, harvest	採草	damage	損傷
cutting frequency	刈取回数，刈取頻度	damage	被害
cutting height	刈取高さ	damping-off, standing dead (material)	立枯れ
cutting interval	刈取間隔		
cutting management	刈取管理		
cutting mechanism	切断機構	dark germination	暗発芽
cutting method	刈取法	dark reaction	暗反応
cutting method	切断法	dark respiration	暗呼吸
cutting stage	刈取ステージ	dark-inhibited seed, light germinator, light sensitive seed	光発芽種子
cutting time, date of cutting	刈取期		
cutting yield	刈取収量		
cyanogenesis	シアン化物生成	data, measured value	測定値
cytochrome oxidase	チトクローム酸化酵素	data logger method	データロガー法
		database	データベース
cytogenetics	細胞遺伝学	date of cutting, cutting time	刈取期
cytokinin	サイトカイニン		

day length, photoperiod	日長
day-neutral plant	中日植物
daytime grazing	昼間放牧
dead material [matter]	枯死部
death	枯死
death [mortality] rate	死亡率
decapitation	頂部除去，断頭
decarboxylase	脱炭酸酵素
deciduous broad-leaved forest	落葉広葉樹林
deciduous forest	落葉樹林
deciduous tree	落葉樹
decline	衰弱，減退，低下
decolorization	脱色
decomposer	分解者
decomposition	分解
decomposition of litter, litter decomposition	リター分解
deep application of fertilizer, deep placement	下層［深層］施肥
deep placement, deep application of fertilizer	下層［深層］施肥
deep plow	深耕プラウ
deep plowing	深耕
deep-rooted crop	深根性作物
defecation	排ふん【糞】
defence	防衛，防御，ディフェンス
deferred grazing	待期放牧
deferred pasture	待期牧区
deferred rotational grazing	待期輪換放牧
deficiency	欠乏
deficiency symptom	欠乏症（状）
definite bud	定芽
defoliated [feeding] damage	食害
defoliation	せん【剪】葉，落葉，摘葉処理
defoliation [browse] tolerance	被食耐性
deforestation, forest clearance	森林伐採
defrosting	解凍
degeneration	退化，変性，逆行変性
degradable protein	分解性タンパク【蛋白】質
degradation, deterioration	荒廃，劣化，衰退
degrading [waste, deteriolated] grassland	衰退［荒廃］草地
degree of freedom	自由度
degree of succession	遷移度
dehydration	脱水，脱水症
dehydration	急速高温度乾燥
dehydrogenase	脱水素酵素，デヒドロゲナーゼ
deionized water	脱塩水，脱イオン水
dejucing	搾汁
delayed germination	発芽遅延
delayed [slow] release fertilizer	緩効性［遅効性］肥料
delignification	脱リグニン
demography	個体群統計学，デモグラフィ，人口学
denaturation	変性
denitrification	脱窒作用
denitrifying [nitrate reducing] bacteria	硝酸還元菌
denitrifying bacteria	脱窒菌
dense [heavier, thick] seeding [sowing]	厚播き

dense [close] planting	密　植	deviation	偏差，偏向
densitometry	デンシトメトリ	diagnosis	診　断
density	密　度	diallel analysis	ダイアレル分析
density dependent	密度依存	diallel crosses	二面交雑
density effect	密度効果		［交配］，
density independent	密度独立		総当たり交雑
deoxyribonucleic acid, DNA	デオキシリボ核酸	dialysis	［交配］ 透析（法）
depositing cattle	預託牛	diameter at breast height, DBH	胸高直径
depositing system	預託方式		
depreciation	減価償却	diarrhea	下　痢
depression	抑制，抑圧	dicotyledon	双子葉植物，
depth of plowing [tillage]	耕起深度 ［耕深］	diet, feed (stuff), fodder	双子葉類 飼料，餌
desaltification	脱塩作用		
desert	砂　漠	diet preference, palatability	採食し【嗜】好性，
desert planting [revegetation]	砂漠緑化		し【嗜】好性
desert revegetation [planting]	砂漠緑化	diet selection	餌選択，飼料選択
desert shrub	砂漠かん【灌】木	dietary selection, selective grazing [herbivory]	選択採食
desertification	砂漠化		
desiccant	乾燥剤	difference	差　異
desiccation, drying, drought	乾　燥	differential host	判別寄主
		differential species	識別種
design of experiment, experimental design	実験計画，試験設計	differentiation	分　化
		diffusion	拡　散
detached leaf	切断葉	diffusion resistance	拡散抵抗
detasseling	雄穂除去	digestibility	消化率，消化性
deteriolated [degrading, waste] grassland	衰退［荒廃］草地	digestible crude protein, DCP	可消化粗タンパク【蛋白】質
deterioration, degradation	荒廃，劣化，衰退	digestible dry matter yield	可消化乾物収量
		digestible energy, DE	可消化エネルギー
determinate inflorescence	有限花序	digestible nutrient	可消化養分，可消化成分
detoxication	解　毒		
development	発育，発生	digestible true protein	可消化純タンパク【蛋白】質
development	開発，発達		
development planning	開発計画	digestion	消化，温浸
developmental phase	発育相	digestion rate	消化速度

digestion trial	消化試験	distribution area	分布地
digestive organ	消化器	distribution type	分布型
digestive tract, alimentary canal	消化管	disturbance	かく【撹】乱
		disturbed habitat [land]	かく【撹】乱地
diluent, dilute solution	希釈剤, 希釈液		
dilute solution, diluent	希釈剤, 希釈液	ditch	みぞ【溝】
diluvial plateau	洪積台地	diurnal range	日較差
diluvial soil	洪積土	diurnal rhythm	日間リズム
diluvium	洪積地（層）	diurnal variation	日変化, 日内変動
dioxin	ダイオキシン		
diploid	二倍体（の）, 複相体（の）	diversified [multiple] farming	複合経営
diplospory	複相胞子生殖	division	分　裂
dipping	薬　浴	DNA marker	DNAマーカー
direct seeding [sowing]	直　播	dolomite	ドロマイト
		domestication	栽培化, 家畜化
direct seeding [sowing] culture	直播栽培	domestication, acclimation, acclimatization	順化, じゅん【馴】化
disappearance rate	消失速度, 消失率		
		dominance	優占, 優位, 優性
disclimax	妨害極相		
discriminant function	判別関数	dominance, abundance	優占度
disease	病　害	dominance effect	優性効果
disease forecasting	病害発生予察	dominant character	優性形質
disease garden	耐病性検定圃	dominant herb [grass]	優占草種
disease resistance [tolerance]	耐病性, 病害抵抗性	dominant species	優占種
		donor	供与体, ドナー
disinfection	消　毒	dormancy	休眠, 休眠性, 休止状態
disk harrow, clod crusher	砕土機, ディスクハロー		
		dormancy awakening	休眠覚醒
		dormancy breaking	休眠打破
disk plow	ディスクプラウ	dormancy form	休眠型
dispersal, dispersion	散布, 分散	dormant [latent] bud	休眠芽
dispersion, dispersal	散布, 分散	dormant period	休眠期間
dispersion map	分散図	dormant seed	休眠種子
dispersion of risk	危険分散	dormant stage, resting period	休眠期
displacement	置換, 排出		
disseminule form	散布器官型	dose	投与量, 服用量, 線量
distillation method	蒸留法		
distortion, malformation	奇　形	double cropping	二期作
		double superphosphate	重過リン酸石灰
distribution	分　布	dough(-ripe) stage	こ【糊】熟期

drainage	排水，排液法，排のう【膿】法	dry matter ratio	乾物率
		dry matter yield	乾物収量
		dry off	乾乳
dressed carcass	枝肉	dry season	乾季
dressing percentage	枝肉歩合	dryer	乾燥機
dried feed	乾燥飼料	drying, desiccation, drought	乾燥
dried skim milk	脱脂粉乳		
drift fence	移動追込柵	drying by drafting, forced-air drying	通風乾燥
drill	条播機，ドリル		
drill (seeding, sowing) width	播き幅	drying by heating	火力乾燥
		drying plant	乾燥施設
drillage	不耕起ドリル播き	drying rate	乾燥速度
		dual purpose breed	兼用種《家畜》
drilling	ドリル播き	dump rake	ダンプレーキ
drinker, guzzler, watering point	飲水施設	dump trailer	ダンプトレーラ
		dung beetle	フン虫《食ふん【糞】性コガネムシ》
drip culture	滴下法		
driveway	誘導路		
driving wheel	駆動輪	dung cart	きゅう【厩】肥車
drought, drying, desiccation	乾燥		
		dung pat	ふん【糞】塊
drought	干[旱]ばつ【魃】	dung patch	不食過繁地，不食地，排ふん【糞】過繁地
drought damage [injury]	干[旱]害		
		dung-deposited place [area], area covered by dung	排ふん【糞】地
drought resistance [tolerance]	耐乾[干，旱]性		
drought stress	乾燥ストレス		
dry (matter) weight	乾（物）重	duplex genotype	複式遺伝子型
dry cow	乾乳牛	duplicate gene	重複遺伝子
dry land farming	乾地農業	durability	耐久性
dry matter	乾物	durability, persistency	永続性
dry matter accumulation	乾物蓄積	durable years, useful life	耐用年数
dry matter digestibility	乾物消化率	duration of sunshine	日照時間
dry matter disappearance rate	乾物消失率	dust coating method	粉衣法
		duster	散粉機
dry matter partitioning ratio	乾物分配率	dwarf	わい【矮】性の，い【萎】縮した
dry matter production	乾物生産，乾物生産量		
		dwarf bamboo pasture, Sasa-type grassland	ササ型草地
dry matter productivity	乾物生産力		

dying	枯死（の）	ecological equivalent	同位種
dynamic strength	力学的強度	ecological optimum	生態的最適性
dynamics	動態	ecological process	生態的プロセス
		ecological pyramid	生態ピラミッド
E		ecological succession	生態遷移
		ecological trait, ecological attribute	生態的特性
ear, head	穂		
ear	雌穂《トウモロコシ》	ecology	生態学
		economic	経済の
ear breaking	切穂	economic [agronomic] character [trait]	経済 [実用, 農業] 形質
ear density	穂密度		
ear emergence, earing, heading	出穂	economical	経済的な
		ecospecies	生態種
ear initiation stage, panicle differentiation stage	幼穂分化期	ecosystem	生態系
		ecotourism	エコツーリズム
		ecotype, biotype	生態型, エコタイプ
ear length	穂長		
ear marking	耳票付け	ecotype selection	生態型選抜
earing, ear emergence, heading	出穂	ectoderm	外はい【胚】葉
		edaphic resources	土壌資源
earliness	早晩性	edge [border] effect	周縁効果
early diagnosis	早期検定	eelworm, nematode	線虫
early generation	初期世代	effect of fertilizer, manuring effect, fertilizer efficiency	肥効
early growth	初期生育		
early harvesting	早刈り		
early heading stage	出穂初期	effective cumulative temperature	有効積算温度
early maturing	早生（の）		
early season culture	早期栽培《水稲》	effective population size	有効集団サイズ
early seeding [sowing]	早播き, 早期播種	effective soil depth	有効土層
		effective temperature	有効温度, 実効温度
early variety	早生品種		
early weaning	早期離乳	efficiency	効率
early-successional species	遷移初期種	efficiency for light energy conversion	光エネルギー転換効率
ear-to-row method	一穂一列法	efficiency of utilization, use efficiency	利用（効）率
easy examination method	簡易検定法		
eating time	採食時間	efflux, run-off, outflow	流出（量）, 流去（量）
ecological attribute, ecological trait	生態的特性		
		egg cell	卵細胞, 卵子
ecological distribution	生態分布		

electric conductivity	電気伝導率, 比電気伝導度
electric fence	電気牧柵
electron transfer system	電子伝達系
electronic capacitance probe, grass [pasture] meter	草量計
electrophoresis	電気泳動（法）
element	要素, 元素
elevated CO_2, CO_2 elevation	CO_2 上昇
elevation, altitude	標高
elongated virus	長形ウイルス
elongation, extension	伸長, 延長
eluviation, leaching	溶脱（作用）
emasculation, castration	除雄《植物》, 去勢《動物》, 性腺摘除《動物》
embryo	はい【胚】, 胎仔
embryo culture	はい【胚】培養
embryogenesis	はい【胚】形成
emergence, budding	出芽, ほう【萌】芽
emergence, appearance	出現
emission	放出, エミッション
enclosure	囲い
end of grazing, housing from pasture, closing the grazing	退牧
endangered (plant) species	絶滅危ぐ【惧】（植物）種
endemic species	固有種, 在来種
endocarp	内果皮
endocrine disruptors	内分泌かく乱化学物質
endodermis	内皮
endogenous	内生（の）
endogenous plant hormone	内生植物ホルモン
endophyte	エンドファイト, 内生菌
endoplasmic reticulum [*pl.* endoplasmic reticula], ER	小胞体
endosperm	はい【胚】乳
endotoxin	内毒素, 菌体内毒素, エンドトキシン
energy balance	エネルギー出納
energy consumption, energy expenditure	エネルギー消費
energy conversion efficiency (rate), energy transfere efficiency, energy exchange rate	エネルギー転換率
energy efficiency	エネルギー効率
energy exchange rate, energy conversion efficiency (rate), energy transfere efficiency	エネルギー転換率
energy expenditure, energy consumption	エネルギー消費
energy metabolism	エネルギー代謝
energy requirement	エネルギー要求
energy transfere efficiency, energy conversion efficiency (rate), energy exchange rate	エネルギー転換率
enforced dormancy	強制休眠
enrichment	富化
ensilage	エンシレージ
ensilage cutter	吹上カッター
ensiling, silage making	サイレージ調製
entomophily, insect pollination	虫ばい【媒】

environment	環境
environment for cultivation	栽培環境
environmental acclimatization	環境順化
environmental assessment	環境アセスメント，環境評価
environmental condition	環境条件
environmental conservation	環境保全
environmental degradation	環境劣化［悪化］
environmental factor	環境要因
environmental impact	環境影響
environmental impact assessment	環境影響評価
environmental monitoring	環境モニタリング
environmental policy	環境政策
environmental pollution	環境汚染
environmental risk assessment	環境リスクアセスメント
environmentally sound agriculture	環境調和型農業
enzyme	酵素，エンザイム
enzyme reaction	酵素反応
epicarp, exocarp	外果皮
epicotyl	上はい【胚】軸
epidemic disease	流行病
epidermal cell	表皮細胞
epidermis	表皮
equilibrium	平衡
equilibrium model	平衡モデル
eradication	撲滅，根絶
erect stem [shoot], orthotropic stem [shoot]	直立茎，垂直茎
erect type	直立型
erosion	浸食《衝撃など による》
erosion	侵食《水，風 などによる》
error	誤差
error variance	誤差分散
escape	野生化
escape response	逃避反応
esophagus	食道
essential amino acid	必須アミノ酸
essential element	必須要素，必須元素
essential fatty acid	必須脂肪酸
established [reclaimed] land	造成地
establishment	造成
establishment	定着
establishment method, method of grassland reclamation	造成工法
estimate, estimated value	推定値
estimated value, estimate	推定値
estimation	推定
estrogen	エストロジェン
estrous	発情，発情期
estrous cycle	性周期，発情周期
ethanol, ethyl alcohol	エタノール，エチルアルコール
ether extract	エーテル抽出物
ethyl alcohol, ethanol	エタノール，エチルアルコール
ethylene	エチレン
etiolation	黄白化
evaluation, assessment	評価
evaluation of nutritive value	栄養評価
evaporation	蒸発
evapotranspiration	蒸発散

evenness	均一性	exponential growth	指数関数的成
evergreen	常緑の		［生］長
evolution	進　化	extension, elongation	伸長，延長
ewe	雌めん【緬】羊	extensive	粗放な
ewe/lamb grazing	母子放牧	extensive	広範な，
	《緬羊》		大規模な
exalbuminous seed	無はい【胚】乳	extensive [rough]	粗放管理
	種子	management	
excess damage	過剰障害	extention	普　及
excess-moisture [humid, wet] injury	湿　害	external feature, external morphology	外部形態
exchangeable acidity	交換酸度	external morphology,	外部形態
exchangeable base	交換性塩基，	external feature	
	置換生塩基	extinction	絶　滅
exclusion experiment	排除実験	extracellular	細胞外（の）
excreta, excretion	排泄物	extract	抽出物［液］
excretion	排泄，排出	extraction	抽出，抽出法
excretion, excreta	排泄物	extremely early maturing	極早生（の）
execution management	施工管理	extremely late	極晩生（の）
execution method	施工法	maturing	
execution standard	施工基準	exudate	分泌物
exocarp, epicarp	外果皮		
exogenous	外生の， 外因性の	F	
exotic species	外来種		
exotic weed	外来雑草	F_1 hybrid	一代雑種，
expectation, expected value	期待値	factor	F_1 雑種 因子，要因
expected value, expectation	期待値	factor analysis factorial analysis	因子分析 要因分析法
expenditure for agricultural management, farm management cost, cost of farming	農業経営費	fall [autumn] crop fall dressing, autumn manuring fall [autumn] seeding [sowing]	秋　作 秋　肥 秋播き
experimental design, design of experiment	実験計画， 試験設計	fallow fallow (land)	休閑（の） 休閑地
experimental field	実験ほ【圃】場	family	家系，系統群
experimental plot	試験区，実験区	family	科
exploitation, searching	探　索	far-red light	近赤外光
exponential distribution	指数散布	farm [business] accounting	経営計算

farm animal, livestock	家畜	reproductivity	種子生産量
farm machinery, agricultural machinery	農業機械	feed additive	飼料添加物
		feed analysis	飼料分析
		feed carrier	飼料運搬車
farm management	農場管理	feed cleaner	飼料精選機
farm management cost, expenditure for agricultural management, cost of farming	農業経営費	feed composition, composition of feed	飼料成分組成
		feed efficiency	飼料効率
		feed flavor	飼料臭
		feed ingredient	配合原料
farm mechanization	農業機械化	feed intake	飼料摂取（量）
farm [business] organization	経営組織	feed mixer	飼料配合機
		feed pelleting machine, pelleting machine	飼料成形機
farm village, rural community	農村，農業集落		
farm wagon	ファームワゴン	feed quality	飼料品質
farm-household mainly engaged in farming	第一種兼業農家	feed requirement	飼料要求量
		feed science	飼料学
		feed (stuff), diet, fodder	飼料，餌
farm-household mainly engaged in other jobs	第二種兼業農家	feedback	フィードバック
		feeder	給餌機
farming machinery	農機具	feeder [stock] cattle	肥育用素牛
farming system	営農体系	feeding	飼養，摂食
farmyard manure	たい【堆】肥	feeding	給与
fasting	絶食，空腹時	feeding [fattening, store] cattle	肥育牛
fat	脂肪		
fat soluble vitamin	脂溶性ビタミン	feeding [defoliated] damage	食害
fat-corrected milk, FCM	脂肪補正乳		
		feeding experiment	飼養試験
fattening	肥育	feeding habit	食性
fattening [feeding, store] cattle	肥育牛	feeding management	飼養管理
		feeding standard	飼養標準
fatty acid	脂肪酸	feeding station	フィーディングステーション
fauna	動物相		
faunal diversity	動物多様性	feeding value	飼料価値
F-distribution	F分布	female [pistillate] flower	雌花
feather meal	羽毛粉，フェザーミール		
		female sterility	雌性不ねん【稔】
feces	ふん【糞】，ふん【糞】便，大便		
		fence	牧柵，柵
fecundity,	繁殖率［力］，	fermentation	発酵

fermentative quality	発酵品質	field capacity	ほ【圃】場容水量
fertile [productive] tiller	有効分げつ	field [upland] crop	畑作物
fertile land	肥よく【沃】地	field crop cultivation	畑作
fertility	ねん【稔】性,繁殖性,受精能	field curing, field drying	ほ【圃】場乾燥
fertility	肥よく【沃】度,肥料分	field drying, field curing	ほ【圃】場乾燥
fertilization	受精	field experiment [test, trial]	ほ【圃】場試験,野外試験
fertilization, fertilizer application, fertilizing	施肥	field husbandry	耕種
		field plot technique	ほ【圃】場試験法
fertilizer, manure	肥料	field-cured hay, sun-cured hay	自然乾草
fertilizer application, fertilizing, fertilization	施肥	filamentous virus	糸状ウイルス
		filling	詰込み,充てん【填】
fertilizer distributor	施肥機		
fertilizer drill	施肥播種機	finger-wheel rake	フィンガホイールレーキ
fertilizer efficiency, effect of fertilizer, manuring effect	肥効	fire	野火
		firebreak	防火帯
		firming, compacting, soil packing	鎮圧
fertilizer injury	こえ[肥]焼け		
fertilizer nutrient	肥料成分	first crop	1番草
fertilizer rate, application rate of fertilizer	施肥量	first cutting	1番刈り
		first cutting hay	1番刈乾草
		first filial generation, F_1	雑種第一代
fertilizer response test	肥効試験		
fertilizer trial	肥料試験	first heading	出穂始め
fertilizing, fertilizer application, fertilization	施肥	first synthetic generation, syn1	合成第1代
		fish meal	魚粉
fertilizing standard	施肥標準	fistula	フィステル,ろう孔[管]
fetus	胎児,胎仔		
fiber	繊維,線維		
fiber degrading [breakdown] enzyme	繊維分解酵素	fitness	適応度
		fixation	固定(化),固定法
fibrous feed	繊維質飼料		
fibrous root	繊維根,ひげ根	fixed assets	固定資産
field	畑(地),ほ【圃】場	fixed capital	固定資本
		fixed cost	固定費用

fixed cutting management	刈取基準	stem bearing a flower bud	
fixed fence	固定牧柵	flowering, anthesis, blooming	開　花
fixed [set] grazing [stocking]	固定放牧, 定置放牧	flowering date	開花日
fixed nitrogen	固定窒素	flowering factor	開花要因
fixed [constant] temperature	定温, 恒温	flowering habit	開花習性
		flowering hormone	花成ホルモン
flag [boot] leaf	止　葉	flowering induction	開花誘起
flail mower	フレールモーア	flowering time, blooming stage	開花期
flail type forage harvester	フレール型フォレージハーベスタ	flower-visiting insect	訪花昆虫
		fluctuation	ほうこう 【彷徨】変異
flake	フレーク		
flame spectrophotometry	炎光分光分析	fluid fertilizer	液状肥料
		fluorescence	蛍　光
flavonoid	フラボノイド	fluorescent X-ray analysis	蛍光X線分析
flavor	風　味		
flint corn	フリントコーン	fluorine, F	フッ素
flooding, waterlogging, submergence	湛水, 洪水	flux	フラックス
		fodder, feed (stuff), diet	飼料, 餌
flora	植物相, フローラ	fodder shrub [tree]	飼料木
floral differentiation, flower bud differentiation	花芽分化	foggage	フォッゲージ, 霜枯れ草
		foggage conservation, stand preservation	立毛貯蔵
floral diversity	植物多様性		
floral zone	植物帯	foliage	茎葉, 葉 《集合的》
florigen	花成素, フロリゲン		
		foliar	葉の, 葉状の, 葉質の
floristic [botanical] composition	種組成《植物》, 植物構成, 草種構成, 草種割合		
		foliar application, foliar spray	葉面散布
		foliar diagnosis	葉面診断
flower	花	foliar residue, residual leaf	残　葉
flower bud differentiation, floral differentiation	花芽分化		
		foliar spray, foliar application	葉面散布
flower head	頭　花	food chain	食物連鎖
flower infection	花器伝染	food cycle	食物環
flower stalk, stem with a flower bud,	着らい【蕾】枝	food web	食物網

foot and mouth disease	口てい【蹄】えき【疫】	formula feed	配合飼料
forage	茎葉飼料	fossil fuels	化石燃料
forage [herbage, pasture] allowance	供給草量, 割り当て草量	foundation seed	原々種
		four-wheel drive tractor	四輪駆動トラクタ
forage blower	フォレージブロワ	fractionation	分画, 分取
		free barn, loose barn	放飼式牛舎, ルースバーン
forage crop	飼料作物		
forage crop field	飼料畑	free breeding on pasture, pasture breeding	まき牛繁殖, 自然交配
forage grass, grass	イネ科(牧)草		
forage harvester	フォレージハーベスタ		
		free choice feeding, cafeteria feeding	自由選択給餌
forage herb	飼料草		
forage [herbage] legume	マメ科(牧)草	free water	自由水, 遊離水
		freeze	凍結
forage production	飼料生産	freeze drying, lyophilization	凍結真空乾燥
forage test	フォレージテスト		
		freezing damage [injury]	凍(結)害
forage wagon	フォレージワゴン		
		freezing hardiness	耐凍性
forage yield	飼草収量	frequency	頻度, 回数
foraging	採餌	frequency curve	度数曲線
forb	広葉草本	frequency dependent	頻度依存
forced-air drying, drying by drafting	通風乾燥	frequency distribution	度数[頻度]分布
forecasting of occurrence	発生予察	frequency of utilization	利用頻度
foreign gene	外来遺伝子	fresh feces	生ふん【糞】, 新鮮ふん【糞】
forest clearance, deforestation	森林伐採		
forest grazing	林内放牧	fresh forage feeding, soiling, zero grazing	青刈り, 生草[青刈]給与
forest soil	森林土壌		
forest understory	林床		
forestland, woodland	樹林地, 林地, 森林	fresh matter	新鮮物
		fresh weight	新鮮重《植物》
forestomach	前胃《反すう【芻】動物》	fresh [green] yield	青刈[生草]収量
forma specialis, f. sp. [*pl.* formae speciales]	分化型	front-end loader	フロントローダ
		frost hardiness, frost resistance [tolerance]	耐霜性
formation	群系		
formic acid	ギ【蟻】酸	frost heaving, heaving	凍上

frost injury	霜害
frost pillar	霜柱
frost resistance [tolerance], frost hardiness	耐霜性
fructan	フルクタン
fructification, seed setting	結実
fructification	子実体（形成）《病》
fructosan	フラクトサン, フルクトサン
fructose	果糖, フラクトース
F-test	F検定
fuel	燃料
full feeding	飽食給与
full heading	穂揃い
full maturity	完熟
full-fermented compost	完熟たい【堆】肥
full-ripe stage	完熟期
full-time household	専業農家
fumigant	くん【燻】蒸剤
function	機能
fungal [fungous] disease	菌類病
fungicide	殺菌剤, 殺真菌薬
fungistasis	静菌作用
fungous [fungal] disease	菌類病
fungus [pl. fungi]	かび類, 菌類, 真菌類
fused magnesium phosphate	よう【熔】（成）リン（肥）

G

galactose	ガラクトース
gall	虫えい, 虫こぶ
gamete	配偶子
gamma ray	γ線
gang plow	多連プラウ
gap	ギャップ, 空げき【隙】
gap size	ギャップの大きさ
gas chromatography	ガスクロマトグラフィ
gas exchange	ガス交換
gaseous phase	気相
gastric juice	胃液
gathering width, swath	刈幅
geitonogamy	隣家受粉
gel	ゲル
gelation	ゲル化
gene	遺伝子
gene analysis	遺伝子分析［解析］
gene expression	遺伝子発現
gene flow	遺伝子流動
gene frequency	遺伝子頻度
gene mutation	遺伝子（突然）変異
gene source	遺伝子源
general combining ability	一般組合せ能力
general purpose tractor	はん【汎】用トラクタ
generation	世代
generation interval	世代間隔
genet	ジェネット
genetic correlation	遺伝相関
genetic diversity	遺伝的多様性
genetic drift	遺伝的浮動［ドリフト］
genetic gain	遺伝改良量［獲得量］
genetic heterogeneity	遺伝的異質性
genetic homogeneity	遺伝的等質性
genetic male sterility	遺伝的雄性不念
genetic resources	遺伝資源
genetic structure	遺伝（的）構造

genetic variability	遺伝的変異性
genetic variation	遺伝的変異
genetics	遺伝学
genome	ゲノム
genome analysis	ゲノム分析
genotype	遺伝子型
genus	属
genus [intergeneric] hybrid	属間（交）雑種
geographic variation	地理的変異
geographical and social condition of location, locational condition	立地条件
geographical distribution	地理的分布
geographical isolation	地理的隔離
geotropism	屈地現象, 屈地性, 向地性
germination	発芽
germination ability, viability of seed	発芽勢
germination habit	発芽習性
germination percentage, percentage of germination	発芽率
germination test	発芽試験
germplasm	生殖質
gestation, pregnancy	妊娠
gibberellin	ジベレリン
glassland plant	草原植物
gley horizon	グライ層
global [total solar, total short wave] radiattion	全天日射, 全短波放射
global warming	地球温暖化
globulin	グロブリン
glucoamylase	グルコアミラーゼ
glucolipid	糖脂質
glucose	グルコース, ブドウ糖
glucoside	配糖体, グルコシド
glume	えい【穎】, 包えい【穎】
glutinous, waxy	もち【糯】性の
glycine betaine	グリシンベタイン
glycolysis	解糖
glycoprotein	糖タンパク【蛋白】質
goat	山羊
golf course	ゴルフ場
good [abundant] harvest	豊作, 上作
goodness of fit	適合度
gradient	こう【勾】配
graftage, grafting	接木
grafting, graftage	接木
grain	穀実, 子実
grain drill	グレーンドリル
grain filling, ripening	登熟
Gramineae	イネ科
gramineous forage crop	イネ科飼料作物
granite	花こう【崗】岩
granitic regosol, Masa	マサ土
granular fertilizer	粒状肥料
grass, forage grass	イネ科（牧）草
grass [pasture] meter, electronic capacitance probe	草量計
grass pasture	イネ科牧草地
grass silage	グラスサイレージ
grass tetany	グラステタニー
grass-grazer mutualism	草・草食獣の共存
grassland	草原, 草地
grassland, sward	草地

grassland [pasture] improvement	草地改良 [改善]
grassland agriculture [farming]	草地農業
grassland community	草地群落
grassland conservation	草地保全
grassland dairying	草地酪農
grassland development	草地開発
grassland diagnosis	草地診断
grassland ecology	草地生態学
grassland ecosystem	草地生態系
grassland farming [husbandry]	草地畜産
grassland forest	牧野樹林
grassland husbandry [farming]	草地畜産
grassland landscape	草地景観
grassland management	草地管理
grassland [pasture] production	草地生産
grassland science	草地学
grassland [pasture] soil	草地土壌
grassland type	草地型
grassland utilization	草地利用
grassland vegetation	草地植生
grassland [pasture] weed	草地雑草
gravel	れき【礫】
gravel culture	れき【礫】耕
gravitational water	重力水
grazed herbage	採食草
grazed sward, grazing land, pasture	放牧地, 放牧草地
grazing, pasture intake	喫食
grazing, stocking	放牧
grazing animal [livestock]	放牧家畜
grazing [ingestive] behavior	採食 [摂取] 行動
grazing behavior	放牧行動
grazing calendar	放牧カレンダー
grazing [carrying, stocking] capacity	牧養力
grazing cattle	放牧牛
grazing cycle	放牧周期
grazing density, stocking density	放牧密度
grazing effect	放牧効果
grazing facility	放牧施設
grazing fee	放牧料
grazing forest [woodland]	混牧林, 牧野林
grazing grass	放牧草
grazing intensity	放牧強度
grazing interval	放牧間隔
grazing land, pasture, grazed sward	放牧地, 放牧草地
grazing livestock [animal]	放牧家畜
grazing management	放牧管理
grazing optimization hypothesis	放牧最適化仮説
grazing period	滞牧日数, 放牧期間
grazing pressure	放牧圧
grazing season	放牧季節
grazing system	放牧法
grazing technique	放牧技術
grazing time	放牧時間
grazing type	牧養型
green fodder, soilage	青刈飼料
green forage	生草
green leaf	緑葉
green manure	緑肥
green [fresh] yield	青刈 [生草] 収量
greenhouse	温室, ガラス室
greenhouse effect	温室効果
greenhouse gases	温室効果ガス
greening	緑変
greeting behavior	挨拶行動
grinding	粉砕, 研磨

gross agricultural income	農業粗収益
gross energy	総エネルギー
gross production	総生産（量）
gross protein value, GPV	タンパク【蛋白】価
ground clearance	地ごしらえ
ground dolomitic, limestone, magnesium-calcium carbonate	苦土石灰
ground substance, matrix, substrate	基質
ground vegetation, vegetation in forest understory	林床植生
ground water level [table]	地下水位
ground water table [level]	地下水位
group feeding	群給餌
growing [cultivated] field	栽培ほ【圃】場
growing period	生育日数
growing point	成［生］長点
growing process	生育過程
growing region	伸長帯
growing season	成［生］長期間, 生育時期, 生育期間
growth	増殖《微生物の》
growth, growth and development	生育, 成［生］長
growth analysis	成［生］長解析
growth and development, growth	生育, 成［生］長
growth cabinet	グロースキャビネット, 植物育成箱
growth curve, increment curve of growth	成［生］長曲線
growth form, growth type	生育型, 成［生］長型, 生育形
growth habit	生育習性, 成［生］長習慣, 生育特性
growth hormone	成［生］長ホルモン
growth inhibitor	成［生］長阻害物質
growth pattern	成［生］長様式
growth period	成［生］長期
growth phase	成［生］長相, 増殖相
growth promotion, growth stimulation	成［生］長促進, 生育促進
growth rate	成［生］長速度
growth regulator	成［生］長調節物質
growth retardant	成［生］長抑制剤［物質］
growth stage	成［生］長段階
growth stimulation, growth promotion	成［生］長促進, 生育促進
growth type, growth form	生育型, 成［生］長型, 生育形
grubbing	抜根
gully erosion	ガリー浸食
gun nozzle	鉄砲ノズル
guzzler, drinker, watering point	飲水施設
gymnosperm	裸子植物

H

habit	習性
habitat	自生地, 生育地, 生［棲］息地
habitat fragmentation	生［棲］息地分断化

habitat segregation, separation	すみわけ	hay crusher	乾草破砕機，ヘイクラッシャ
hail injury	ひょう【雹】害		
hair	被毛	hay cube	ヘイキューブ
hairy root	毛根，毛状根	hay dryer	牧草乾燥機
halophyte	塩生植物	hay elevator	ヘイエレベータ
hand cutting	手刈り	hay fork, pitch fork	ヘイフォーク
handling, operation	操作，ハンドリング	hay grade	乾草規格
		hay lift	乾草揚げ機，ヘイリフト
haploid	半数体，単相体，一倍体	hay loader	乾草積込機，ヘイローダ
haprocorm	球茎《チモシー》	hay making	乾草調製
hard seed	硬実	hay press, stationary hay baler	乾草圧縮機，ヘイプレス
hardening	硬化，ハードニング		
		hay production	乾草生産
hardness	硬度	hay quality	乾草品質
harrow	ハロー	hay rake	ヘイレーキ
harrowing	は【耙】耕，ハローイング	hay stacker	乾草積上機，ヘイスタッカ
harvest, mechanical (forage) harvesting, cutting	採草	hay tedder	乾草反転機，ヘイテッダ
		hay value	乾草価
harvest	収穫(する)	haylage	ヘイレージ
harvest method	収穫法	hazardous waste	有害廃棄物
harvest time	収穫期	head, ear	穂
harvester	収穫機，ハーベスタ	head of grazing animals	放牧頭数
harvesting date	収穫日	heading, ear emergence, earing	出穂
hastening of germination	催芽，発芽促進		
		heading date [stage, time]	出穂期
haulm, stem, stalk	茎		
hay	乾草	headland	枕地
hay baler	乾草こん【梱】包機，ヘイベーラ	heart rate	心拍数
		heat, hot temperature	暑熱
		heat budget [balance]	熱収支
hay baling	乾草こん【梱】包	heat [thermal] capacity	熱量
hay carrier	乾草運搬機	heat resistance [tolerance]	耐暑性，耐熱性
hay conditioner	乾草圧砕機，ヘイコンディショナ		
		heat transfer	熱伝導

heated-air dryer	熱風乾燥機，火力乾燥機
heated-air drying	熱風乾燥，加熱乾燥
heathland	ヒース（の茂る荒れ地）
heating treatment	加熱処理
heaving, frost heaving	凍上
heavy clay	重しょく【埴】土
heavy grazing [stocking]	重放牧
heavy manuring culture	多肥栽培
heavy metal	重金属
heavy metal pollution	重金属汚染
heavy nitrogen, ^{15}N	重窒素
heavy soil	重粘土
heifer	未経産雌牛，若雌牛
heliophilous [lightfavored] seed, light germinating seed	好光性種子
heliotropism	屈日性
helium, He	ヘリウム
hemicellulose	ヘミセルロース
hemicryptohytes	半地中植物
hemoglobin	ヘモグロビン，血色素
herb, herbaceous plant	草本
herb species	草種
herbaceous plant, herb	草本
herbage	草《茎葉》
herbage [forage, pasture] allowance	供給草量，割り当て草量
herbage consumption	被食
herbage intake, intake	採食（量）
herbage [forage] legume	マメ科（牧）草
herbage mass	草量
herbage [pasture] plant	牧草
herbicide	除草剤
herbicide tolerance [resistance]	除草剤耐性
herbivore, herbivorous animal	草食動物
herbivorous animal, herbivore	草食動物
herd	群《動物》
herd management	群管理
herd size	群の大きさ
heredity, inheritance	遺伝
heritability	遺伝率［力］
hermaphrodite	両性花
heterogeneity	異質性，異種性，不均一［質］性
heterogenesis, alternation of generations	世代交代
heterokaryon	異核接合体［共存体］
heteroploid, aneuploid	異数体
heteroploidy, aneuploidy	異数性
heterosis, hybrid vigor	ヘテローシス，雑種強勢
heterotroph	従属栄養生物，従属栄養体，有機栄養生物
heterotrophism, heterotrophy	従属栄養
heterotrophy, heterotrophism	従属栄養
heterozygote	異型［異種，ヘテロ］接合体
hexose	六炭糖
hibernation, wintering	冬眠
high altitude cool-climate region	高冷地

H [125]

high and stable production	高位安定生産	chromosome homozygote	同型［ホモ］接合体
high level cutting	高刈り	hoof cultivation	てい【蹄】耕法
high moisture silage	高水分サイレージ	horizontal distribution	水平分布
high order tiller, higher-nodal-position tiller	高次［位］分げつ	horse	馬
		host	宿主，ホスト
		host range	宿主範囲［域］
high producing dairy cow	高泌乳牛	host-specific toxin	宿主特異的毒素
		hot temperature, heat	暑熱
high productivity	高位生産性	housing	舎飼
high ridge	高うね【畦】	housing from pasture, end of grazing, closing the grazing	退牧
high temperature injury	高温障害		
high temperature stress	高温ストレス	hull, husk, chaff	もみ【籾】殻
		humic acid	腐植酸
high yield	多収	humid region	多雨地帯
high yielding ability	多収性	humidification	加湿
high yielding culture	多収（穫）栽培	humidity	湿度，湿気
high-analysis mixed fertilizer	高度化成肥料	hummock	凸状草の
		humus	腐植（質）
higher-nodal-position tiller, high order tiller	高次［位］分げつ	hunting	狩猟
		husk, chaff, hull	もみ【籾】殻
		husking	もみ【籾】すり
hill, ridge	うね【畝，畦】	hybrid, crossbred	交雑種，雑種
hill seeding [sowing], spaced planting	点播	hybrid vigor, heterosis	雑種強勢，ヘテローシス
hillside grassland, sloping pasture	傾斜草地	hybridization, crossing	交雑，雑種形成
		hybridization [cross] breeding	交雑育種
hillside [sloping] meadow	傾斜採草地		
		hydraulic system	油圧装置
hilly grassland	高原草地	hydrochloric acid	塩酸
histochemical	組織化学的	hydrocyanic acid	青酸
histological	組織学的	hydrogen ion concentration	水素イオン濃度
histology	組織学		
Holstein	ホルスタイン（種の）	hydrogen, H	水素
		hydrolase	加水分解酵素
homeostasis	生体恒常性，ホメオスタシス	hydrolysis	加水分解
		hydrophyte	水生植物
		hydrophytic [aquatic] weed	水生雑草
homogenization	均質化		
homologous	相同染色体		

hydroponics, solution [water] culture	水耕	imperfect state	不完全時代, 無性時代
hygroscopicity	吸湿性	impermeable layer	不透性層
hypersensitivity	過敏性, 過感受性, 過敏症	impermeable seed	不吸湿性種子
		improved pasture	改良草地, 改良放牧地
		improved variety	改良品種
hypha [pl. hyphae], mycelium [pl. mycelia]	菌糸体, 菌糸	improvement	改良
		improvement of grassland	草生改良
hypocotyl	下はい【胚】軸, はい【胚】軸	in vitro	インビトロ, 試験管内の, 生体外の
hypocotyl culture	はい【胚】軸培養		
		in vitro culture	試験管内培養, インビトロ培養
hypomagnesaemia	低マグネシウム血症		
hypothesis	仮説, 仮定	in vitro [artificial] digestibility	人工消化率

I

		in vivo	インビボ, 生体内の
identification	同定, 識別	inactivation	不活性化, 失活
ideoecology, autoecology	個生態学	inbred line	近交系（統）, インブレッドライン
idling	アイドリング, 遊歩《家畜の》		
		inbreeding	近親交配[近交], 同系交配
ill-drained paddy field	湿田		
illuminance, light intensity	光強度, 照度	inbreeding	同系繁殖, 内交配
image analysis	画像解析	inbreeding coefficient	近交係数
image processing	画像処理	inbreeding depression	近交弱勢[退化]
imago, adult	成(個)体, 成虫		
immature	未成熟の, 未熟な	inclination, slope [slant] angle	傾斜（角）度
		income	所得, 収入
immature barnyard manure, incomplete manure	未熟たい【堆】肥	incompatibility	不親和性, 不和合性, 不適合性
immature embryo	未熟はい【胚】	incompatibility	配合禁忌, 配合変化
immunity	免えき【疫】(性)		
imperfect fungi	不完全菌		

incomplete manure, immature barnyard manure	未熟たい【堆】肥
increasing CO_2 concentration	CO_2 濃度増加
increment curve of growth, growth curve	成[生]長曲線
incubation	定温培養，インキュベーション
incubation [latent] period	潜伏期間
indefinite [indeterminate] inflorescence	無限花序
index	指数
index method	指示物質法
indicator	指標物質，指標
indicator plant	指標植物，植物指標
indigenous, native, spontaneous	自生（の），在来（の）
indigenous [local, native] variety	在来品種，在来種，在来系統
indirect effect	間接効果
individual	個体
individual density	個体密度《生物》
individual identification	個体識別
individual selection	個体選抜
individual variation	個体変異
indole acetic acid, IAA	インドール酢酸
induced dormancy	誘発休眠
induction	誘発，誘導
inductively coupled plasma [ICP] emission spectrometry	誘導結合プラズマ[ICP]発光分光分析
infection, affection	感染
infection, transmission	伝染
infertile land	不毛地
infertility, sterility	不ねん【稔】性
inflorescence	花序
inflow, influx	流入（量）
influx, inflow	流入（量）
infrared gas analyzer	赤外線ガス分析計
ingestion	摂取，経口摂取
ingestive [grazing] behavior	採食[摂取]行動
ingredient	成分
ingredient composition	飼料組成
inheritance, heredity	遺伝
inherited character	遺伝形質
inhibiting [repressor] gene, inhibitor	抑制遺伝子
inhibition	阻害
inhibitor	阻害剤，インヒビター
inhibitor, inhibiting [repressor] gene	抑制遺伝子
inhibitory factor	抑制因子
injury by continuous cropping	連作（障）害
injury of regrowth	再生障害
innate dormancy	内生休眠
inoculation	接種
inoculation feeding	接種吸汁
inoculum [pl. inocula]	接種源
inorganic component, mineral	無機成分
insect	昆虫
insect damage [injury]	虫害
insect fauna	昆虫相
insect pest, pest insect	害虫
insect pollination, entomophily	虫ばい【媒】
insect resistance [tolerance]	耐虫性
insect vector	ばい【媒】介昆虫

insecticide	殺虫剤	interspecific	種間（の）
insemination	授精	interspecific competition	種間競争
insoluble carbohydrate	不溶性炭水化物	interspecific competition of grass, competition of grass	草種競合
inspection	検査，検診		
intake	摂取量		
intake, herbage intake	採食（量）		
integrated control	総合防除	interspecific hybrid	種間雑種
integrated solar radiation	積算日射量	interspecific hybridization	種間交雑
intensive cultivation	集約栽培	interspecific variation	種間変異
intensive farming	集約的農業	intertillage	中耕
intensive grazing	集約放牧	intraspecific	種内の
interaction	相互作用，交互作用（項）	intraspecific competition	種内競争
		introduced crop	導入作物
interaction, interrelationship	相互関係	introduced genetic resources	導入遺伝資源
intercellular	細胞間の	introduced species	導入（草）種
intercellular space	細胞間げき【隙】	introduced variety	導入品種
		introduction	導入
intercellular substance	細胞間物質	introduction breeding	導入育種
intercepted light intensity, received light intensity	受光量	invader	侵入者
		invading [alien] plant	侵入植物
		invertebrate	無脊椎動物
inter-cropping	間作	investigation of actual condition, actual condition survey	実態調査
interference	干渉		
intergeneric crossing	属間交雑		
intergeneric [genus] hybrid	属間（交）雑種	investment	投資
		iodine, I	ヨウ素
intermediary metabolism	中間代謝	ion-electrode method	イオン電極法
		ion-exchange resin	イオン交換樹脂
intermediate [alternate] host	中間宿主	iron, Fe	鉄
		irradiation	照射
intermediate product	中間生成物	irrigated grassland	かんがい【灌漑】草地
intermittent irrigation	間断かんがい【灌漑】	irrigation	かんがい【灌漑】
internal distribution	体内分布		
internode	節間（部）	irrigation requirement	用水量
internode elongation	節間伸長	isoelectric point	等電点
interrelationship, interaction	相互関係	isoenzyme, isozyme	アイソザイム，イソ[同位]酵素
interrow space	株間		

isolate	分離株	kurtosis	せん【尖】度
isolated culture	隔離栽培		
isolation	隔　離		**L**
isolation	単　離		
isolation field	隔離ほ【圃】場	labor expense	労働費
isotonic solution	等張液	labor intensity	労働集約度
isotope	アイソトープ, 同位体[元素]	labor productivity	労働生産性
		labor saving	省力（の）, 省力化
isozyme, isoenzyme	アイソザイム, イソ[同位]酵素	labor saving management	省力管理
		laboratory test	室内検定
	J	lactating [milking, dairy] cow	搾乳牛, 泌乳牛
		lactation	泌乳, 授乳, 乳汁分泌
Japanese Beef cattle	和　牛		
Japanese Black cattle	黒毛和種牛	lactation curve	泌乳曲線
Japanese Brown cattle	褐毛和種牛	lactation experiment	泌乳試験
Japanese Polled cattle	無角和牛	lactation period	泌乳期間, 授乳期
Japanese Shorthorn	日本短角種		
jaw movement	あご運動	lactic acid	乳　酸
Jersey	ジャージー牛	lactic acid bacteria	乳酸菌
joint soil	目　土	lactic fermentation	乳酸発酵
judgement of feed	飼料鑑定	lactose	乳　糖
juiciness	多汁性	lag phase	静止期, 誘導期
		lagoon	ラグーン, 沼
	K	LAI [leaf area index]	葉面積指数
		lamb production	子めん【緬】羊生産
karyotype	核　型		
kernel depth [size]	粒　径	lamina [pl. laminae], leaf blade	葉　身
kernel size [depth]	粒　径		
ketosis	ケトーシス, ケトン症	land classification	土地分類
		land improvement	土地改良
key species	主要種	land leveler	地ならし機
keystone species	キーストーン種	land [soil] preparation, leveling of ground	整　地
kind of crop	作　目		
kind of land	地　目	land productivity	土地生産性
kinetin	カイネチン, キネチン	land reclamation	干拓, 土地造成
		land rent	地　代
Kjeldahl method	ケルダール法	land restoration	土地回復[修復]
Knop's solution	クノップ液		
Krebs cycle	クレブス回路	land roller, packer	鎮圧ローラ

land utilization [use]	土地利用	law of minimum	最小律
land value	土地評価	law of segregation	分離の法則
LANDSAT	ランドサット	lawn, turf	芝(地)，芝生
landscape	ランドスケープ，景観	lax [rough] grazing	粗放牧
		layer mixing plow	混層耕プラウ
landscape design	景観設計	leaching	浸出，洗脱，
landscape simulation	景観シミュレーション		リーチング
		leaching, eluviation	溶脱(作用)
LAR [leaf area ratio]	葉面積比	lead, Pb	鉛
large implement	大農具	leader-follower grazing	先行後追放牧
large-size farm machinery	大型農機具	leading variety	基幹品種
large-size power farming system	大型機械化体系	leaf	葉
		leaf analysis	葉分析
larva [pl. larvae]	幼虫，幼生	leaf appearance rate	出葉速度
laser	レーザ	leaf area	葉面積
last frost	終　霜	leaf axil	葉えき【腋】
late cutting	晩刈り	leaf base	葉基［脚］
late frost	晩　霜	leaf blade, lamina [pl. laminae]	葉　身
late maturation	晩　熟		
late maturing	晩生(の)	leaf conductance	葉コンダクタンス
late maturity	晩熟性		
late seeding [sowing]	晩播，遅播き	leaf emergence [appearance]	出　葉
late variety	晩生品種		
latent [dormant] bud	休眠芽	leaf emergence [appearance] interval	出葉間隔
latent heat	潜　熱		
latent infection	潜在［潜伏］感染		
		leaf extension rate	葉身伸長速度
latent [incubation] period	潜伏期間	leaf longevity	葉の寿命
		leaf position on stem	葉　位
lateral bud	側　芽	leaf sheath, sheath	葉しょう【鞘】
lateral root	側　根	leaf vein, nerve, rib	葉　脈
lateral shoot	側　枝	leafiness	多葉性
late-successional species	遷移後期種	leaflet	小　葉
		leaf-stem ratio	葉茎比
latin squares design	ラテン方格法	lean meat	赤身肉
lattice design	格子配列法	learning	学　習
laurel forest zone	照葉樹林帯	least significant difference, LSD	最小有意差
law of constant final yield	最終収量一定の法則		
		legume, pod	豆　果
law of diminishing returns	収量漸減の法則	Leguminosae	マメ科
		leguminous	マメ科の

leguminous [root nodule] bacteria	根粒菌	light rhythm, photoperiodism	光周期
lemma	外えい【穎】	light saturation	光飽和
length of frost free season	無霜期間	light sensitive seed, dark-inhibited seed, light germinator	光発芽種子
lenient [light] grazing	軽度放牧		
lesion	病はん【斑】, 病変	light transmission	光透過率
		light-distribution [intercepting] characteristics	受光態勢
lethal concentration	致死濃度		
lethal does 50, LD_{50}	半数致死量		
lethal dose	致死量	light-inhibited seed	嫌光性種子
level	水準	light-receiving [intercepting] coefficient	受光係数
level of significance, significant level	有意水準		
leveling of ground, land [soil] preparation	整地	lightfavored [heliophilous] seed, light germinating seed	好光性種子
ley	輪作草地, 短年輪作草地	lignification	木化, リグニン化
ley farming	穀草式輪作		
ley pasture	短年放牧地	lignin	リグニン
life cycle	生活環	ligule	葉舌
life form	生活形［型］	lime	石灰
life history	生活史	lime sower	石灰散布機
light clay	軽しょく【埴】土	limestone	石灰岩
		liming material, calcium fertilizer	石灰質肥料［資材］
light compensation point	光補償点		
		limiting amino acid	制限アミノ酸
light energy	光エネルギー	limiting factor	限定［規制］要因
light extinction coefficient	吸光係数		
		limiting resource	制限的資源
light germinating seed, heliophilous [lightfavored] seed	好光性種子	line [pedigree] selection	系統選抜
		line interception method	ラインインターセプション法
light germination	光発芽, 明発芽		
light germinator, dark-inhibited seed, light sensitive seed	光発芽種子	line transect method	ライントランセクト法
		linear model	線形模型
light [lenient] grazing	軽度放牧	linear programing	線形計画
light intensity, illuminance	光強度, 照度	linear regression	直線回帰
		linearity	直線性
light reaction	明反応		

lipase	リパーゼ, 脂肪分解酵素	location, site	立地
lipid	脂質	locational condition, geographical and social condition of location	立地条件
liquid fertilizer [manure]	液(体)肥(料)		
liquid manure distributor	液肥散布機	locus	遺伝子座, 座位
		lodging	倒伏
liquid nitrogen	液体窒素	lodging control	倒伏防止
liquid phase	液相	lodging resistance	耐倒伏性, 倒伏抵抗性
litter, bedding	敷料, 敷わら【藁】		
		logarithmic scale	対数目盛
litter	遺体, 落葉・落枝	logarithmic transformation	対数変換
litter decomposition, decomposition of litter	リター分解	logisitic formula [equation]	ロジスチック式
		logistic curve	ロジスチック曲線
litter layer	リター層		
live weight	生体重《動物》	logistic equation [formula]	ロジスチック式
live [body] weight gain	増体, 体重増加		
		long-day plant	長日植物
livestock, farm animal	家畜	long-day treatment	長日処理
livestock exclusion, prohibition of grazing	禁牧	longevity	寿命, 長寿
		long-lived seed	長命種子
		long-range forecast	長期予報
livestock farm, ranch	牧場	long-term prediction	長期予測
livestock farming, animal husbandry	畜産, 畜産経営	long-term research	長期研究
		loose baler	ルースベーラ
livestock [animal] industry	畜産業	loose barn, free barn	放飼式牛舎, ルースバーン
livestock waste, animal manure	家畜ふん【糞】尿	loss	損失
		lost [missing] plant	欠株
livestock-grazing regime	家畜放牧管理	lot	ロット
		Lotka-Volterra equation	ロトカ・ボルテラ式
loader	ローダ		
loam	壌土	low cost	低コスト
loamy sand	壌質砂土	low temperature germinability	低温発芽性
local [native, indigenous] variety	在来品種, 在来種, 在来系統		
		low temperature injury	低温障害
local climate	局地気候	low temperature seed vernalization	低温種子春化処理
local lesion	局部病はん【斑】		
		lower layer	下層
locality	地域性, 局所性	lower leaf	下葉

lower order tiller, lower-nodal-position tiller	低次[位]分げつ	main product	主産物
		main [tap] root	主根
		main stem	主茎
lower-nodal-position tiller, lower order tiller	低次[位]分げつ	maintenance behavior	維持行動《採食，休憩などの》
lowland	低地	maintenance ration	維持飼料
lowland grassland	低地草原	major element, macroelement, macronutrient	多量要素[元素]
low-level cutting	低刈り		
low-moisture silage	低水分サイレージ	major gene resistance	主働遺伝子抵抗性
low-nutrient environment	低養分環境	male [staminate] flower	雄花
lying	横が【臥】		
lyophilization, freeze drying	凍結真空乾燥	male sterility	雄性不ねん【稔】(性)
lysigenous aerenchyma	破生通気組織	malformation, distortion	奇形
lysimeter	ライシメータ	malnutrition	栄養失調(症)，栄養不良
lysis	溶菌，溶解《細胞などの》	maltose	麦芽糖，マルトース
M		mammary gland	乳せん【腺】
		management, care	管理
macro seed-pellet	マクロシードペレット	management [business] balance	経営収支
macroclimate	大気候	manganese, Mn	マンガン
macroconidium [pl. macroconidia]	大型分生子，大型分生胞子	manifestation, phenotypic expression	形質発現
macroelement, macronutrient, major element	多量元素[要素]	mannan	マンナン
		mannose	マンノース
macronutrient, macroelement, major element	多量要素[元素]	manure, stable [barnyard] manure	畜肥，きゅう【厩】肥，肥[こえ]
magnesium, Mg	マグネシウム	manure, fertilizer	肥料
magnesium-calcium carbonate, ground dolomitic limestone	苦土石灰	manure loader	たい【堆】肥積込機
		manure management	ふん【糞】尿処理
main crop	主作物		
main culm	主かん【稈】	manure spreader	たい【堆】肥散布機
main effect	主効果，主作用		

[134] M

manuring effect, effect of fertilizer, fertilizer efficiency	肥効
manuring practice	肥培管理
mapping	地図化，マッピング
marker	マーカー
marking method	標識法
Markovian succession model	マルコフ遷移モデル
marsh	湿地，沼地
Masa, granitic regosol	マサ土
masking	病徴隠ぺい，不顕性の，しゃへい【遮蔽】
mass [bulk] emasculation	集団除雄
mass seed production	集団採種
mass selection	集団選抜
mast, nut	木の実
mastication, chewing	そしゃく【咀嚼】
mastitis	乳房炎, 乳せん【腺】炎
mat formation	マット形成
material, substance, matter	物質
material [matter] cycle, circulation of material	物質循環
maternal line	母系 (系統)
maternal line selection	母系選抜
mating	交配，接合，交尾
mating type	交配 [接合] 型
matrix, ground substance, substrate	基質
matrix model	行列モデル
matter, substance, material	物質
matter production	物質生産
maturation, maturity, ageing	熟成
maturation	成熟
mature embryo	成熟はい【胚】
mature field	熟畑
mature plant	成熟個体《植物》
maturing stage	熟期
maturity, maturation, ageing	熟成
maturity [ripening] stage	成熟期
maximum dry matter production rate	最大乾物生産速度
maximum moisture [water] capacity	最大容水量
maximum tiller number stage	最高分げつ期
meadow	採草地
meadow soil	低湿地土
meal	荒粉，ひき割り
mean	平均 (値)
mean square	平均平方
mean temperature	平均温度
means	手段
measured value, data	測定値
measurement	測定
measuring instrument	計測装置
mechanical (forage) harvesting, cutting, harvest	採草
mechanical property	機械特性
mechanization	機械化
mechanized working	機械作業
median	中央値
medium-maturing	中生 (の)
meiosis	減数 [成熟, 還元] 分裂
mercury, Hg	水銀
meristem	分裂組織, 成 [生] 長点

M [135]

mesh data	メッシュデータ
mesh figure [map]	メッシュ図
mesocotyl	中はい【胚】軸
mesophyll	葉肉
messenger RNA, m-RNA	メッセンジャーRNA
metabolic live weight	代謝体重
metabolism	代謝
metabolite	代謝(産)物, 代謝生成物
metabolizable energy, ME	代謝エネルギー
metalloprotein	金属(結合)タンパク【蛋白】質
meteorological [climatic] element	気候要素
meteorological index of production	気候生産指数
meteorological productivity	気象生産力
meteorology	気象
methemoglobin, MHb	メトヘモグロビン
methemoglobinemia	メトヘモグロビン血液
method of (grassland) establishment	造成法
method of fertilizer application	施肥法
method of grassland reclamation, establishment method	造成工法
method of least squares	最小自[二]乗法
microbody	マイクロボディ, 微小体
microclimate	微気候
microconidimu [pl. microconidia]	小型分生子, 小型分生胞子
microelement,	微量元素
micronutrient, minor element, trace element	[要素]
microenvironment, microsite	微細環境, 微小環境
microflora	微生物相, ミクロフローラ
micronutrient, microelement, minor element, trace element	微量元素, [要素]
microorganism	微生物
microplot	マイクロプロット
microsite, microenvironment	微細環境, 微小環境
microsome	ミクロゾーム《微粒体》, マイクロソーム
microtopography	微地形
middle lamella	中層, 中葉
midrib	中央脈, 中肋
migrule form	繁殖型
milk	牛乳
milk composition, milk ingredient	乳成分
milk fat	乳脂肪
milk fat percentage, butter fat percentage	乳脂率
milk ingredient, milk composition	乳成分
milk [dairy] performance	泌乳能力
milk producing yield	泌乳量
milk production	産乳
milk quality	乳質
milk replacer	代用乳
milk yield	乳量
milker	搾乳機, ミルカ
milking [lactating, dairy] cow	搾乳牛, 泌乳牛

milking curve	搾乳曲線	mixed [companion] planting	混植
milk-ripe	乳熟		
milk-ripe stage	乳熟期	mixed silage	混合サイレージ
millet	雑穀	mixed sown pasture	混播放牧草地
mineral	無機物, ミネラル	mixed sown sward	混播草地
		mixture	混合 (剤)
mineral, inorganic component	無機成分	mixture	混在
		mixture rate [ratio]	混生率 [割合]
		mode of distribution, pattern of spatial distribution	分布様式
mineral deficiency	要素欠乏		
mineral fertilizer	無機質肥料		
mineral nutrition	無機栄養	mode of inheritance	遺伝様式
mineral soil	鉱質土壌	model	モデル
mineral uptake	ミネラル吸収	modeling	モデリング
minimum effective temperature	最低有効温度	modular plant	モジュール植物
		module	モジュール
minimum fatal low temperature	最低致死温度	moisture	水分
		moisture [water] capacity	容水量
minimum tillage	ミニマムティレッジ		
		moisture [water] content	含水量, 水分含量
minor element, microelement, micronutrient, trace element	微量元素 [要素]		
		moisture equivalent	水分当量
		molasses	糖蜜
		mold [mould]	かび
Miscanthus-type [silvergrass-type] grassland	ススキ型草地	mold inhibitor	抗かび剤
		moldboard plow	はつ土板プラウ
		molding, covering with soil	覆土
missing [lost] plant	欠株		
missing plot [value]	欠測値	moldy smell	かび臭
missing value [plot]	欠測値	mole	モグラ
mitochondrion [pl. mitochondria]	ミトコンドリア	mole drain	弾丸暗きょ【渠】
		mole drainer	モグラ暗きょ【渠】せん孔機
mitosis	有糸分裂		
mix seeding [sowing]	混播, 混ぜ播き		
mixed [multiple] cropping	混作		
		molecular weight	分子量
mixed farming	混合農業	molybdenum, Mo	モリブデン
mixed [blended] fertilizer, composed manure	配合肥料	monitoring	モニタリング
		monitoring equipment	モニタリング機器
mixed grazing	混 (合放) 牧	mono- and oligo-saccharide	単少糖
mixed hay	混合乾草		

monocotyledon	単子葉植物	mulching cultivation, mulch culture	マルチ栽培, 被覆栽培
monogastric animal	単胃動物	multiple alleles	複対立遺伝子
monosaccharide	単糖（類）	multiple [mixed] cropping	混作
moor, bog	湿原	multiple cropping	多毛作
morphogenesis	形態形成［発生］	multiple cross [crossing]	多系交雑
morphology	形態（学）	multiple crossing [cross]	多系交雑
mortality	死亡	multiple [diversified] farming	複合経営
mortality [death] rate	死亡率	multiple infection	重複［多重］感染
mother plant	母株, 母本	multiple landuse	多面的土地利用
mother [original, parent] tiller	母茎	multiple regression	重回帰
mother tree, seed bearer	母樹	multiplication	増殖
motor cell	機動細胞	multiplier effect	相乗効果
mottle	はん【斑】紋	multivariate analysis	多変量解析
mountain dairy	山地酪農	mutant	突然変異体, 変異体, 変異株
mountain grassland	山間草地	mutation	突然変異
mountain [alpine] grazing	山地放牧	mutual shading	相互しゃへい【遮蔽】
mountain village	山村	mutualism, symbiosis	共生, 相利共生
mountainous area	山間地帯, 山岳地帯, 山地	mycelium [pl. mycelia], hypha [pl. hyphae]	菌糸体, 菌糸
mounting	乗が【駕】	mycorrhiza	菌根
movable cage	移動ケージ	mycotoxin	マイコトキシン, かび毒
movable fence	移動牧柵		
mower	刈取機, モーア		
mower conditioner	モーアコンディショナ		
mowing	草刈り		
muck soil	黒泥土		
mucopolysaccharide	ムコ多糖		
mulberry field [plantation]	桑園		
mulberry plantation [field]	桑園		
mulch	被覆, マルチ		
mulch culture, mulching cultivation	マルチ栽培, 被覆栽培		
mulching	敷草, けい【畦】面被覆, マルチング		

N

N,P,K elements, three primary nutrients, three major nutrients	肥料三要素
national forest	国有林
national park	国立公園
national reserve	国立保護区

native, indigenous, spontaneous	自生（の）, 在来（の）	necrotic lesion	えそ【壊疽】 [え【壊】死] 病はん【斑】
native grassland [pasture], rangeland	野草地	nectar	花　蜜
native herb	自生野草	neighbor	近隣個体
native pasture, wild grassland used for grazing	野草放牧地	nematocide	殺線虫剤, 抗線虫薬
		nematode, eelworm	線　虫
native [local, indigenous] variety	在来品種, 在来種, 在来系統	nematode disease	線虫病
		nerve, leaf vein, rib	葉　脈
		nest quadrat method	重ねワク法
natural community	自然群落	net assimilation rate, NAR	純同化率
natural crossing	自然交雑		
natural curing	自然乾燥	net [apparent] photosynthesis	純光合成, 見かけの光合成
natural enemy	天　敵		
natural forest [stand]	天然林		
natural grassland	自然草地	net production	純生産
natural hybrid	自然雑種	netted vein	網状脈
natural mutation	自然突然変異	neutral detergent fiber, NDF	中性デタージェント繊維
natural pollination	自然受粉		
natural population	自然集団	neutralization	中　和
natural regeneration [reproduction]	天然更新	N-fixation, nitrogen fixation	窒素固定
natural reseeding, volunteering	自然下種	niche	ニッチ, ニッチェ
natural seeding	天然下種	nickel, Ni	ニッケル
natural seedling	自然実生, 天然苗	nicotinamide adenine dinucleotide phosphate, NADP	ニコチンアマイドアデニンジヌクレオチドリン酸
natural selection	自然とうた 【淘汰】, 自然選択		
		nicotinamide adenine dinucleotide, NAD	ニコチンアマイドアデニンジヌクレオチド
natural supply	天然供給（量）		
naturalized species	帰化種		
naturalized weed	帰化雑草	nighttime grazing	夜間放牧
nature conservation	自然保全 [保護]	nitrate	硝酸塩
		nitrate nitrogen	硝酸態窒素
nature reserve	自然保護区	nitrate plant	硝酸植物
near-infrared reflectance spectroscopy	近赤外分光法	nitrate poisoning	硝酸（塩）中毒
		nitrate reductase	硝酸還元酵素
		nitrate reduction	硝酸還元
necrosis	え【壊】死, ネクローシス	nitric acid	硝　酸
		nitrification	硝化作用

nitrifier, nitrifying bacteria	硝化菌 [硝酸化成菌]	element	
		non-capillary pore	非毛管孔げき【隙】
nitrifying bacteria, nitrifier	硝化菌 [硝酸化成菌]	non-equilibrium model	非平衡モデル
nitrite	亜硝酸塩, 亜硝酸の	non-grazing, rest	休牧
		non-grazing [rest] period	休牧期間
nitrite forming bacteria	亜硝酸菌	non-productive tiller	無効分げつ
nitrite nitrogen	亜硝酸態窒素	non-protein nitrogen, NPN	非タンパク【蛋白】態窒素
nitrite oxidizing bacteria	硝酸菌		
nitrate reducing [denitrifying] bacteria	硝酸還元菌	non-seasonal	時なし性《周年出穂性》
		non-structural carbohydrate, NSC	非構造性炭水化物
nitrogen assimilation	窒素同化（作用）	non-tilled cropping, plowless farming, zerotillage	不耕起栽培
nitrogen cycle	窒素サイクル		
nitrogen excess	窒素過剰		
nitrogen fixation, N-fixation	窒素固定	non-tilled establishment, unplowed grssland establishment	不耕起草地造成
nitrogen free extract, NFE	可溶無窒素物		
nitrogen metabolism	窒素代謝	normal crop	平年作
nitrogen source	窒素源	normal curve	正規曲線
nitrogen starvation	窒素きが【飢餓】	normal deviation	正規偏差
		normal distribution	正規分布
nitrogen, N	窒素	normal population	正規母集団
nitrogeneous compound	窒素化合物	no-rooted node	非発根節
		no-tillage	不耕起
nitrous acid	亜硝酸	noxious weed	有害雑草
nodal root	節根	nuclear membrane, septum	核膜
node	節		
node density	節密度	nucleic acid	核酸
node order	節位	nucleus	核
node position	着生節位	number of days for germination	発芽日数
nomadic pastoralism, nomadism	遊牧		
		number of ears [panicles]	穂数
nomadism, nomadic pastoralism	遊牧		
		number of individuals, population size	個体数
nomads	遊牧民		
non-available [unavailable]	不可給態要素	number of stems [tillers]	茎数

number of tillers [stems]	茎　数	on-and-off [rationed] grazing	時間制限放牧
nurse crop	保護作物	one crop system, single cropping	一毛作
nurse trees	保護樹林《林業》	open cattle shed	追込牛舎
nursery [seedling] test	幼苗[幼植物]検定	open pollination	放任受[授]粉, 開放受[授]粉
nursing, suckling	ほ【哺】乳		
nut, mast	木の実	operating accuracy, quality of work	作業精度
nutrient	栄養素, 栄養分		
nutrient deficiency	栄養素欠乏	operation, handling	操作, ハンドリング
nutrient stress	栄養[養分]ストレス	operation [work] system	作業体系
nutrient uptake [absorption]	養分吸収	optimal clover content	最適クローバ率
nutrient-use efficiency	栄養素利用効率	optimal diet	最適餌
nutrition	栄養	optimal foraging	最適採餌
nutritive component	栄養成分	optimization	最適化
nutritive ratio	栄養比	optimum leaf area index, opt. LAI	最適葉面積指数
nutritive value	栄養価		
nylon bag method	ナイロンバッグ法	optimum period	適　期
		optimum range	好適範囲
nymph	若　虫	optimum stage for cutting	刈取適期

O

		optimum temperature	最適温度
		organ	器　官
obligate parasite	純寄生者[菌], 絶対寄生菌	organella	細胞(小)器官, オルガネラ
offset harrow	オフセットハロー	organic acid	有機酸
		organic composition	有機組成
offspring, progeny	後代, 子孫, 次代	organic agriculture [farming]	有機農業[農法]
old [aged] grassland, aging pasture [meadow]	経年草地	organic fertilizer	有機質肥料
		organic matter	有機物
		origin, place of origin	原産地
olfaction	きゅう【嗅】感覚, 臭覚	original breed	原種《家畜》
		original [stock] seed	原種《作物》, もと種
oligosaccharide	オリゴ糖, 少糖		
oligotrophic	栄養不足の[不良の], 貧栄養の	original [mother, parent] tiller	母　茎
omasum	第三胃		

orthotropic stem [shoot], erect stem [shoot]	直立茎，垂直茎
osmosis	浸透現象
osmotic pressure	浸透圧
outbreeding, outcrossing, allogamy	他配，他殖性
outcrossing, outbreeding, allogamy	他配，他殖性
outflow, run-off, efflux	流出（量），流去（量）
ovarian cyst	卵巣のう腫
over luxuriant [rank] growth	過繁茂
overall application	全面散布
overall treatment	全面処理《除草剤の》
overcompensation	過剰調節
overdominance	超優性
overgrazing	過放牧
overhead flooding, submergence	冠水
over-population	個体数過剰
over-seeding [sowing], reseeding	追播き，追播
over-seeding culture	追播栽培
overwintering, wintering	越冬
overwintering bud	越冬芽
ovule	はい【胚】珠
ovule culture	はい【胚】珠培養
ox	去勢雄牛《性成熟後》
oxalic acid	シュウ酸
oxidase	酸化酵素
oxidation	酸化
oxidation-reduction potential, redox potential	酸化還元電位，Eh
oxidative phosphorylation	酸化的リン酸化（反応）
oxygen evolution	酸素放出
oxygen uptake	酸素吸収
oxygen, O	酸素
ozone	オゾン
ozone layer	オゾン層

P

packer, land roller	鎮圧ローラ
paddock, stock yard, yard	追込場
paddock	牧区
paddy field	水田
paddy field and upland rotation, rotation between paddy and dry field	田畑輪換
paddy field in paddy-upland rotation	輪換田
paddy soil	水田土壌
palatability, diet preference	採食し【嗜】好性，し【嗜】好性
palea	内えい【穎】
palisade tissue	柵状組織
pan-breaker	心土破砕機
panicle	円すい【錐】花序，穂
panicle [ear, head, inflorescence] type	穂型
panicle [young ear] formation stage	幼穂形成期
panicle differentiation stage, ear initiation stage	幼穂分化期
panmyelopathia	はん【汎】骨ずいろう【髄癆】

parasite	寄生者［体］, 寄生虫, 寄生生物
parasitism, parasitization	寄生
parasitization, parasitism	寄生
parenchyma	柔組織
parent [mother, original] tiller	母茎
parent-offspring correlation	親子相関
part	部位
parthenogenesis	処女［単為］生殖
partial blooming	部分開花
partial correlation	偏相関
partial dominance	部分優性
partial plowing	部分耕起
partial regression	偏回帰
partial tillage	部分耕うん【耘】
partial tillage culture	部分耕栽培
particle gun	パーティクルガン
partition, allocation	分配, 配分
parts per million, ppm	百万分率
part-time farm household, sidework farmer	兼業農家
parturition, calving	分娩
passive absorption	受動的吸収
pastoral association	牧野組合
pasturage right	放牧権
pasture, grazing land, grazed sward	放牧地, 放牧草地
pasture [forage, herbage] allowance	供給草量, 割り当て草量
pasture breeding, free breeding on pasture	まき牛繁殖, 自然交配
pasture establishment	草地造成
pasture harrow	パスチャハロー
pasture intake, grazing	喫食
pasture [grass] meter, electronic capacitance plobe	草量計
pasture of deposit cattle	預託牧場
pasture [herbage] plant	牧草
pasture plant for short-term use	短年用牧草
pasture [grassland] production	草地生産
pasture renovation	草地更新
pasture [grassland] soil	草地土壌
pasture [grassland] weed	草地雑草
patch	過繁地, パッチ
patch [spot] grazing	パッチ状採食
patch model	パッチモデル
patchiness	パッチ状
patchy distribution	パッチ状分布
pathogen, causal agent	病因, 病原（体）
pathogenic fungus	病原菌
pathogenicity, virulence	病原力［性］
pathotype	病原型, 寄生型
pattern of spatial distribution, mode of distribution	分布様式
PCR [polymerase chain reaction] method	PCR法
peat	泥炭, ピート
peat soil	泥炭土
pectic acid	ペクチン酸
pectin	ペクチン
pedicel	小花［果］柄, 小枝こう【梗】
pedigree	血統, 系図

pedigree breeding [method]	系統育種法	permanent quadrat	固定コドラート
pedigree method [breeding]	系統育種法	permanent transect	固定トランセクト
pedigree [line] selection	系統選抜	permanent wilting	永久しおれ
pedology, soil science	土壌学	permeability	透過性, 浸透性, 透過率
pedometer	歩数記録装置, 万歩計	peroxidase	パーオキシダーゼ
peduncle	花柄, 花梗	persistency, durability	永続性
peeling treatment	はく皮処理	persistency	生存年限
pellet	ペレット, 固形飼料	pest	有害生物
		pest control	害虫防除
pelleting machine, feed pelleting machine	飼料成形機	pest insect, insect pest	害虫
		pesticide, agricultural chemical	農薬
pen barn	追込式畜舎		
penetration	浸透	petiole	葉柄
pentose	五炭糖	pF-soil moisture curve	pF-水分曲線
percentage of germination, germination percentage	発芽率	pF-value	pF値
		phanerophytes	地上植物
		phenol, phenolic acid	フェノール酸
		phenolic acid, phenol	フェノール酸
percentage of leaves	葉部割合	phenolic substance [compound]	フェノール性物質
percentage of ripening	ねん【稔】実歩合［率］		
		phenology	生物季節学, フェノロジー
perennation organ	越冬［永年生］器官		
		phenotype	表現型［形］
perennial	宿根性, 多年生(の)	phenotypic expression, manifestation	形質発現
perfect state	完全［有性］時代《カビ》	phenotypic plasticity	表現型可塑［変］性
performance, productivity	生産力［性］	phloem	師［篩］部
		phosphate absorption coefficient	リン酸吸収係数
performance test	能力［性能］検査		
		phosphate fertilizer	リン酸質肥料
performance test	生産力検定	phosphoenol pyruvic acid, PEP	フォスフォエノールピルビン酸
performance test	性能試験		
periodic occurrence	周期的発生		
periodicity	周期性, 周期現象	phosphorus, P	リン
permanent grassland	永年(牧)草地		

phosphorylase	フォスフォリラーゼ, 過リン酸分解酵素
photochemical reaction	光化学反応
photodecomposition, photolysis	光分解
photolysis, photodecomposition	光分解
photomorphogenesis	光形態形成
photoperiod, day length	日長
photoperiodic control	日長調節
photoperiodic response	光周［日長］反応
photoperiodism, light rhythm	光周期
photorespiration	光呼吸
photosensitivity, sensitivity to daylength	感光性
photosynthate, photosynthetic product	光合成産物
photosynthesis	光合成
photosynthesis of a community	群落光合成
photosynthetic capacity	光合成能（力）
photosynthetic efficiency	光合成効率
photosynthetic organ	光合成器官
photosynthetic product, photosynthate	光合成産物
photosynthetic rate	光合成速度
photosynthetic tissue	光合成組織
phototaxis	走光性
phototropism	屈光性
phyllody	葉化
phyllotaxis	葉序
phylogenetic variation	系統（間）変異
physical	物理的
physiochemical	生理化学的
physioecological	生理生態学的
physiognomy	相観
physiographic factor	地形要因
physiological	生理学的
physiological characteristics	生理的特性
physiological disease	生理病
physiological disorder	生理障害
physiological integration	生理的統合
physiological specialization	生理的分化, 寄生性分化
physiologically [potentially] basic fertilizer	生理的塩基性肥料
phytochrome	ファイトクローム, フィトクロム, フィトクローム
phytohormone, plant hormone	植物ホルモン
phytophagous insect	食植性昆虫
phytoplasma	ファイトプラズマ, マイコプラズマ様微生物
phytotoxicity	植物毒性, 殺草性
phytotron	ファイトトロン, 人工気象室《植物用》
pick up [self-loading] forage wagon	ピックアップワゴン
pick up unit	ピックアップユニット
pig	豚
pigment	色素

pilot communal livestock farm, pilot livestock ranch for common use	共同利用模範牧場	plant nutrient	植物養分
		plant nutrition	植物栄養
		plant performance	個体 [植物体] の成績
pilot livestock ranch for common use, pilot communal livestock farm	共同利用模範牧場	plant physiology	植物生理学
		plant pigment	植物色素
		plant shape	植物体の形状
		plant size	個体サイズ《植物》
pipe duster	パイプダスタ		
piroplasmosis	ピロプラズマ病	plant sociology	植物社会学
pistil	雌ずい，めしべ	plant spacing, planting distance	栽植 [植付] 間隔
pitch fork, hay fork	ヘイフォーク		
place of origin, origin	原産地	plant succession	植生遷移
plain	平原	plant tissue	植物組織
plant, stand	株	plant turnover	植物の交替 [入替わり]
plant	植物		
plant anatomy	植物解剖学	plant type	草型
plant body	植物体	plant variety protection	植物品種保護
plant colonization	植物のコロナイゼーション [すみつき]		
		plant vigor	草勢
		plant weight	個体重《植物》
plant community	植物群落	planted area	作付面積
plant composition	植物組成	planting	植付け，定植，栽植
plant cover [coverage]	植被		
plant demography	植物個体群統計学	planting density	栽植密度
		planting distance, plant spacing	栽植間隔，植付間隔
plant density	個体密度《植物》		
		planting furrow	植え溝
plant distribution pattern	植物分布様式	planting opener	作条器
		planting pattern	栽植様式 [方式]
plant ecology	植物生態学		
plant functional type	植物の機能型 [タイプ]	plasma membrane	原形質膜
		plastic film	プラスチックフィルム
plant [sward] height	草高		
plant hormone, phytohormone	植物ホルモン	plastic limit	塑性限界
		plasticity	可塑性
plant husbandry, cultivation	栽培管理	plastid	色素体，プラスチド
plant husbandry [cultivation] method	栽培法	Pleioblastus-type grassland	アズマネザサ型 [ネザサ型] 草地
plant length	草丈		
plant morphology	植物形態学	plow	すき

plow sole	耕盤		微動遺伝子
plowing	耕起	polymorphism	多型 (現象)
plowing [tilling] method	耕起法	polyploid	倍数体, 多数性
		polyploidization	倍数化
plowing and harrowing, tillage, cultivation	耕うん【耘】	polyploidy	倍数性
		polyploidy breeding	倍数性育種
		polysaccharide	多糖, 多糖体
plowless farming, non-tilled cropping, zerotillage	不耕起栽培	population	個体群, 集団, 母集団
		population density	個体数［群］密度, 生［棲］息密度
pneumatic system	通気組織		
pod, legume	豆果		
pod, shell	さや【莢】		
Podzol	ポドゾル土	population dynamics	個体群動態
poikilothermal	変温性の	population genetics	集団遺伝学
point quadrat method	ポイントコドラート法	population improvement	集団改良
poisoning	中毒	population mean	母平均
poisonous plant	有毒植物	population size, number of individuals	個体数
polder	干拓地		
pollen	花粉		
pollen dispersal	花粉飛散	population structure	集団［個体群］構造
pollen fertility	花粉ねん【稔】性		
		porosity	孔げき【隙】率, 空げき【隙】率
pollen mother cell	花粉母細胞		
pollen parent	花粉親		
pollen tube	花粉管	portable grass meter, portable pasture meter	携帯用草量計
pollination	送粉, 受粉, 花粉ばい【媒】介		
		portable pasture meter, portable grass meter	携帯用草量計
pollinator	送粉者, ポリネータ, 花粉ばい【媒】介者		
		possible duration of sunshine	可照日数
		post-grazing pasture [herbage] mass, residual herbage (mass)	残草 (量)
pollinator	花粉ばい【媒】介昆虫		
pollutant	汚染物質		
polluted water, sewage	汚水	potassium fertilizer	カリ肥料
pollution	汚染	potassium, K	カリウム
polycross	多交配	potential growth	最大可能成［生］長量
polygene	ポリジーン, 多遺伝子,		

potential soil fertility [productivity]	潜在地力
potentially [physiologically] basic fertilizer	生理的塩基性肥料
power farming system	機械化体系
power take-off shaft, PTO	動力取出し軸
power-driven trailer	駆動トレーラ
-3/2 power law of natural thinning	-3/2乗則《自然間引きの》
practical use	実用性
prairie	プレーリー《北米中部の内陸草原》
preceding crop	前作物
preceding cropping	前作
precipitation, amount of precipitation	降水（量）
preconditioning, accustoming, acclimation	じゅん【馴】致
precursor	前駆［先駆］物質［体］
precursor forage plant	前駆牧草
prediction	予測
prediction model	予測モデル
preference ranking	選択順位
pregnancy, gestation	妊娠
prehension	捕捉《草の》
preservation, conservation, storage	貯蔵，保存
preservative, antiseptic	防腐剤
press wheel	鎮圧輪
pretreatment	前処理［処置］
prevention	予防
prewilting, wilting	予乾
primary dormancy	一次休眠
primary production	一次生産
primary root	一次根
primary succession	一次遷移
primary tiller	一次分げつ
primordium [pl. primordia]	原基
principal component	主成分
problem plant	問題植物
processed [compressed] hay	成形乾草
processing	加工
producer	生産者，産生株
production	生産，生成
production	生産量
production cost, productive expenses	生産費
production efficiency	生産効率
production energy	生産エネルギー
production plan	生産計画
production technique	生産技術
productive expenses, production cost	生産費
productive [fertile] tiller	有効分げつ
productivity, performance	生産力，生産性
profit and loss method	損益計算
progeny, offspring	後代，子孫，次代
progeny test	後代検定
progressive succession	進行遷移
prohibition of grazing, livestock exclusion	禁牧
proline	プロリン
prop root	支持根
propagation	増殖《植物》
proper use	適正利用
prophase	前期，分裂前期
prostrate, creeping, stoloniferous	ほふく【匍匐】型（の）
protandry	雄ずい先熟
protease	タンパク【蛋白】質分解酵素，プロテアーゼ

protect cage	保護ケージ
protected area	保護地域
protection	保　護
protection forest	保安林
protein	タンパク【蛋白】質
protein efficiency ratio, PER	タンパク【蛋白】効率
protein nitrogen	タンパク【蛋白】態窒素
protein production	タンパク【蛋白】合成
protein production rate	タンパク【蛋白】合成能
protein supplement	タンパク【蛋白】質飼料
proteolysis	タンパク【蛋白】質分解
protogyny	雌ずい先熟
protoplasm	原形質
protoplast	原形質体，プロトプラスト
protozoa	原生動物，原虫，プロトゾア
pubescence, pubescent	軟毛（でおおわれている）
pubescent, pubescence	軟毛（でおおわれている）
public [cooperative] grassland	公共草地
public [cooperative] pasture, public [cooperative] stock farm	公共牧場
public pasture for raising cattle	公共育成牧場
public [cooperative] stock farm, public [cooperative] pasture	公共牧場
pulling	引き抜き
pupa	さなぎ【蛹】
purchased feed	購入飼料
purchased price	購入価格
pure line	純　系
pure line selection	純系選抜
pure stand	単播草地
purification	浄　化
purification treatment	浄化処理
purity	純度，精選度
put-and-take stocking	頭数調整放牧
putrefaction	腐　敗
pycnidiospore	柄胞子
pycnidium [pl. pycnidia]	柄子殻
pyrenoid	ピレノイド，核様体
pyridoxin, vitamin B_6	ピリドキシン ビタミン B_6
pyruvic acid	ピルビン酸

Q

quadrat	コドラート，方形コドラート，方形区
quadrat method	方形区法，区画法，枠法
quadrat sampling	枠抽出
quadratic curve	二次曲線
quadrivalent [tetravalent] chromosome	四価染色体
qualitative analysis	定性分析
qualitative character	質的形質
quality	品質，質
quality of work, operating accuracy	作業精度
quantitative analysis	定量分析

quantitative character	量的形質	rancher	牧場経営者
quantitative genetics	量的遺伝学	ranching	牧場経営
quantitative inheritance	量的遺伝		《農業類型》
		random arrangement	無作為配置
quarantine	検えき【疫】, 防えき【疫】	random distribution	ランダム［不規則, 機会的］分布
Quaternary	第四紀層		
quick acting manure, readily available fertilizer	速効性肥料	random drift	機会の変動
		random sampling	無作為抽出
		randomization	無作為化
quick lime	生石灰	randomized block design	乱塊法

R

		range	牧野《米国の自然草地》
race	菌系, レース	rangeland, native grassland [pasture]	野草地
race, species	種		
raceme	総状花序	RAPD [random amplified polymorphic DNA] analysis	RAPD分析
rachis, cob	穂軸		
radial vascular bundle	放射維管束		
radiation	放射, 放射線, 照射		
		rapeseed meal	ナタネかす【粕】
radiation breeding	放射線育種		
radicle	幼根, 小根	rapid rotation [short time rotation, short duration] grazing	短期輪換放牧
radicoid form	地下器官型		
radioactive isotope	放射性同位元素		
radioautograph	ラジオオートグラフ		
		rate of feed cost to milk sold	乳飼比
radiotelemetry	ラジオテレメトリ		
		rate of increase	増加速度（率）
rain damage	雨害	rate of work	作業能率
rain gauge	雨量計	ratio of legumes	マメ科率
rainfall	降雨	rationed [on-and-off] grazing	時間制限放牧
rainfall apparatus	降雨装置		
rainfall erosion	降雨侵食	raw milk	生乳
rainfall intensity	降雨強度	ray	放射組織
rainy season	梅雨, 雨季	reaction	反作用, 反応
raising cattle	育成牛	readily available fertilizer, quick acting manure	速効性肥料
raising farm	育成牧場		
rake	レーキ		
ram	雄めん【緬】羊	reaping against lodging	向刈り
ramet	ラメット		
ranch, livestock farm	牧場		

rearing [raising] of seedling	育苗
rearing on pasture	放牧育成
received light intensity, intercepted light intensity	受光量
recessive character	劣性形質
recessive gene	劣性遺伝子
reciprocal crossing	正逆交雑
reclaimed land	開墾地，開拓地
reclaimed [established] land	造成地
reclamation	開墾，開拓
recolonization	再侵入
recombinant	組換型
recombination value	組換価
recommended variety	奨励品種
recording thermometer, thermograph	自記温度計
recovery	回収（率），回復，歩留り
recreation	レクリエーション
recruitment	加入
recruitment pattern	加入様式［パターン］
rectal palpation	直腸検査
recurrence [return] of estrous	発情回帰
recurrent selection	循環選抜
redox potential, oxdation-reduction potential	酸化還元電位，Eh
reducing power, reduction ability	還元力
reducing sugar	還元糖
reduction	還元
reduction ability, reducing power	還元力
red-yellow soil	赤黄色土
reflectance, reflectivity	反射（率）
reflectivity, reflectance	反射（率）
regeneration, renovation	更新
regeneration, reproduction	再生産
regeneration cutting [felling]	更新伐
registered feed	登録飼料
registered seed	登録種子
registered variety	登録品種
registration	登録
registration of plant varieties for the Seeds and Seedlings Law	種苗登録
regosol	未熟土
regression	回帰
regression	退行
regression analysis	回帰分析
regression coefficient	回帰係数
regression curve	回帰曲線
regression line	回帰直線
regrowth	再成[生]長，再生
regrowth vigor	再生草勢
regular distribution	規則分布
regularity	規則性
regurgitation	吐戻し，逆流，弁閉鎖不全
related [allied] species	近縁種
relative growth, allometry	相対成[生]長
relative growth rate, RGR	相対成[生]長率
relative humidity	相対湿度
relative light intensity	相対照度
relative turgidity	水分飽和度
relative weight	相対重量
release effect	放飼効果

remaining crop after harvesting	刈残し
remote sensing	リモートセンシング
removal	除去
renovated pasture [grassland]	更新草地
renovation, regeneration	更新
reorganization on paddy field use	水田利用再編
repellent	忌避剤，忌避物質，駆散薬
replacement	置換，入替え
replacement stock	更新用家畜
repressor [inhibiting] gene, inhibitor	抑制遺伝子
reproduction, breeding	繁殖《動物》，生殖《動物》
reproduction, regeneration	再生産
reproductive allocation	繁殖分配
reproductive [breeding] cattle	繁殖牛
reproductive difficulties	繁殖障害
reproductive effort	繁殖投資，繁殖効率
reproductive part	繁殖器官
reproductive stage	生殖成[生]長期，繁殖期
reproductive tiller [stem]	繁殖茎
reproductivity, fecundity	繁殖率[力]，種子生産量
requirement	要求(量)
reseeding, over-seeding [sowing]	追播き，追播
reserve [storage] organ	貯蔵器官
reserve substance	貯蔵物質
reserve [storage] tissue	貯蔵組織
residual herbage (mass), post-grazing pasture [herbage] mass	残草(量)
residual leaf, foliar residue	残葉
residual root, root residue	残根
residue	残さ【渣】，残留物
resistance, tolerance	抵抗性，耐性，抵抗力
resistant variety	抵抗性品種
resource	資源
resource allocation	資源分配，資源配分
resource partitioning	資源分割
resowing	再播種
respiration	呼吸(作用)
respiration activity	呼吸能
respiratory enzyme	呼吸酵素
respiratory quotient, RQ	呼吸係数
respiratory substance	呼吸基質
rest, non-grazing	休牧
rest [non-grazing] period	休牧期間
resting period, dormant stage	休眠期
resting place	たてば，休息場
restricted [controlled] feeding	制限給餌
retention	滞留
retention time	滞留時間
reticulorumen	第一・二胃，反すう【芻】胃
reticulum	第二胃，細網《組織》

retinol, vitamin A_1	レチノール, ビタミン A_1
retrogressive succession	退行遷移
return of animal excreta	ふん【糞】尿還元
return [recurrence] of estrous	発情回帰
revegetation	緑化
RFLP [restriction fragment length polymorphism]	制限酵素断片長多型
rhizomatous plant	根茎植物
rhizomatous type	根茎型
rhizome, subterranean stem	地下茎, 根茎
rhizosphere	根圏
rib, leaf vein, nerve	葉脈
rib	主脈《植物》
riboflavin, vitamin B_2	リボフラビン, ビタミン B_2
rice bran	米ぬか
rice cultivation [culture]	稲作
rice culture [cultivation]	稲作
rice straw	稲わら【藁】
rice terrace, terraced paddy field	棚田
ridge, hill	うね【畝, 畦】
ridge breadth	うね【畦】幅
ridge culture	うね【畦】立て栽培
ridge making	うね【畦】立て
rights of common	入会権
rill erosion	細流浸食
riparian grassland	河岸草地
ripening, grain filling	登熟
ripening	ねん【稔】実
ripening period	登熟期間
ripening [maturity] stage	成熟期
ripening stage	登熟期
river terrace	河川敷
rodent	げっ歯類, 野ねずみ
rodenticide	殺そ剤
roll bale	ロールベール
roll baler, round baler	ロールベーラ, ラウンドベーラ
root	根
root apex	根端
root cap	根冠
root crops	根菜類
root exudate	根の分泌物
root hair	根毛
root knot nematode	根こぶ線虫
root mat	ルートマット
root nodulation	根粒形成
root nodule	根粒
root nodule [leguminous] bacteria	根粒菌
root pressure	根圧
root pruning	せん【剪】根
root residue, residual root	残根
root rot	根ぐされ
root spread	根ばり
root system	根系, 根群
root tuber	塊根
root zone	根域
root/shoot ratio	地下部/地上部比
rooted node	発根節
rooting	発根
rootlet	細根
rosette	そう【叢】生, ロゼット
rosette plant	ロゼット植物
rotary cultivation [tilling]	ロータリ耕
rotary drum mower	ドラムモーア

rotary mower	ロータリモーア		消化
rotary tedder	ジャイロ形テッダレーキ	rumen exercise	ルーメン運動
		rumen fermentation	ルーメン［第一胃］内発酵
rotary tiller	ロータリ耕うん【耘】機		
rotary tilling [cultivation]	ロータリ耕	rumen fluid	第一胃内溶液
		rumen liquid	ルーメン液
rotation, crop rotation	輪作	rumen protozoa	第一胃内原生動物
rotation between paddy and dry field, paddy field and upland rotation	田畑輪換		
		ruminant	反すう【芻】動物［家畜］，反すう【芻】の
rotation grazing, rotational grazing	輪換放牧		
		rumination	反すう【芻】
rotational grazing, rotation grazing	輪換放牧	rumination behavior	反すう【芻】行動
rotational upland field	転換畑	rumination time	反すう【芻】時間
rough [lax] grazing	粗放牧		
rough [extensive] management	粗放管理	ruminology	ルミノロジー
		runner, creeper, stolon	ほふく【匍匐】茎
rough [simple] renovation	簡易更新		
		runner	ランナー
roughage, bulky feed	粗飼料	run-off, efflux, outflow	流出（量），流去（量）
round bale silage	ロールベールサイレージ		
		run-off loss	流亡
round baler, roll baler	ロールベーラ，ラウンドベーラ	rural amenity	田園アメニティ
		rural area	農村地域
		rural community, farm village	農村，農業集落
row [stripe] seeding [sowing]	条播		
		S	
row planting	列植え		
row space	うね【畦】間	saccharide	糖類
rugose	縮葉		
rumen	第一胃，反すう【芻】胃，ルーメン	saccharose, sucrose	ショ糖，スクロース
		saline soil	塩類土壌
rumen bacteria	第一胃［ルーメン］内細菌	saline water	塩（性）水，生理食塩水
		salinity	塩分
rumen bypass	ルーメンバイパス	salinization	塩（性，類）化
		saliva	だ【唾】液
rumen digestion	ルーメン（内）	salt, sodium chloride	食塩，塩化

	ナトリウム	screening	スクリーニング
salt accumulation	塩類集積	scrub, shrub	低［かん】【灌】
salt damage [injury]	濃度障害, 塩類障害	scutellum	木林 はい【胚】盤,
salt resistance [tolerance]	耐塩性	search time	小板, 盤状体 探索時間
salts	塩　類	searching, exploitation	探　索
sample	標本, 試料, 検体	seasonal breeding seasonal change [variation]	季節繁殖 季節変動
sample mean	標本平均	seasonal form	季節型
sampling	試料採取, 標本抽出	seasonal grazing seasonal productivity	季節放牧 季節生産性
sampling error	抽出誤差	seasonal succession	季節遷移
sand	砂　土	seasonal variation	季節変動
sand culture	砂　耕	[change]	
sand dune [hill]	砂丘（地）		
sand hill [dune]	砂丘（地）	seasonal variation	季節変異
sandstone	砂　岩	seasonality	季節性
sandy loam	砂壌土	second flush	2番芽
sandy soil	砂質土壌	second flush	2番草
sap	汁　液	secondary cell wall	二次細胞壁
sap inoculation	汁液接種	secondary [winter] cropping	裏　作
sap transmission	汁液伝染		
sapling	幼令木［樹］	secondary forest	二次林
saponin	サポニン	secondary growth	二次成【生】長
saprophyte	腐生者［菌］, 雑菌	secondary metabolite secondary production	二次代謝物 二次生産
Sasa-type grassland, dwarf bamboo pasture	ササ型草地	secondary root secondary succession secondary tiller	二次根 二次遷移 二次分げつ
satoyama	里　山	secretion	分　泌
saturated fatty acid	飽和脂肪酸	sedimentary rock	たい【堆】積岩
saturation	飽　和	seed	種子, 種［たね］
savannah	サバンナ 《熱帯の草原》	seed bank seed bearer, mother tree	種子バンク 母　樹
scan width	スキャン幅		
scarification	種皮［刺傷］ 処理	seed [seeding, sowing] bed	播種床
science of soil and plant nutrition	土壌肥料学	seed bed preparation seed certification	播種床造成 種子証明 [検定]
sclerotium [pl. sclerotia]	菌　核	seed clean	種子精選

seed coat	種皮	seed-borne disease	種子伝染病
seed coating	種子被覆	seeder, seeding [sowing] machine	播種機
seed collection	種子収集		
seed disinfection	種子消毒	seeding, sowing	播種，たねまき
seed dispersal	種子散布	seeding [sowing, seed] bed	播種床
seed dressing	種子粉衣		
seed fertility	種実[種子]ねん【稔】性	seeding [sowing] date	播種日
		seeding [sowing] depth	播種深（度）
seed furrow [stripe]	播きみぞ【溝】		
seed germination	種子発芽	seeding [sowing] machine, seeder	播種機
seed grading	選種		
seed identification	種子鑑別	seeding [sowing] method	播種法
seed inoculation	種子接種		
seed inspection [testing]	種子検査	seeding [sowing] rate	播種密度，播種量
seed multiplication	種子増殖	seeding [sowing] time	播種（時）期
seed output	種子量	seedling	実生，芽ばえ
seed parent	種子親	seedling	幼苗，幼植物
seed pellet	シードペレット	seedling age	苗令
		seedling establishment	実生［幼植物］の定着
seed pelleting	種子造粒		
seed preservation	種子保存	seedling selection	実生選抜
seed presoaking	浸種	seedling stage	幼苗[稚苗]期
seed pretreatment	種子前処理	seedling [nursery] test	幼苗[幼植物]検定
seed production	採種（栽培），種子生産		
		seeds and seedlings	種苗
seed propagation	種子繁殖	segregation	分離
seed quality	種子品質	selection	選抜
seed quality control	種子品質管理	selection by specific gravity	比重選
seed set percentage	結実率		
seed setting, fructification	結実	selection criterion	選抜基準
		selection effectiveness	選抜効果
seed size	種子サイズ	selection method	選抜法
seed standards	種子分類基準	selective absorption	選択吸収
seed storage	種子貯蔵	selective grazing [herbivory], dietary selection	選択採食
seed testing [inspection]	種子検査		
seed transmission	種子伝染	selective herbicide	選択性除草剤
seed treatment	種子処理	selective herbivory [grazing], dietary selection	選択採食
seed vernalization	種子春化		
seed weight	種子重		
seed yield	種子収量	selective medium	選択培地

self-compatibility	自家和合性	sequential grazing	時間差放牧
self-feeder	セルフフィーダ	seral stage	途中段階
self-fertilization, selfing, autogamy	自家受精, 自殖	serine	《遷移の》 セリン
self-formulation	自家配合	serpentinite	蛇紋岩
self-incompatibility	自家不和合性	set [fixed] grazing [stocking]	固定[定置] 放牧
self-loading [pick up] forage wagon	ピックアップ ワゴン	setting up the grazing, turning out to pasture	入 牧
self-pollinated crop	自家受粉 [自殖性] 作物	sewage, polluted water	汚 水
self-pollination	自家受粉	sewage sludge	下水汚泥
self-sterility	自家不ねん 【稔】性	sewage sludge fertilizer	汚泥肥料
self-supplying feed	自給飼料	sex	性
self-supplying manure	自給肥料	sexual reproduction	有性生殖
self-thinning	自己間引き	shade, shelter	ひ【庇】蔭
selfed line	自殖系統	shade [cheese] cloth	寒冷紗
selfing, self-fertilization, autogamy	自家受精, 自殖	shade leaf	陰 葉
		shade tolerance	耐陰性
		shade tree [beares]	陰 樹
semi-arid	半乾燥(の)	shading	遮 光
semi-arid region	半乾燥地	shading culture	ひ【庇】陰栽培
seminal root	種子根	shading pasture plant	ひ【庇】陰牧草
semi-natural grassland	半自然草地 [草原]	shale	ケツ【頁】岩
		shallow tillage	浅 耕
semipermeable membrane	半透膜	shattering	脱 粒
		shattering habit	脱粒性
senescence, ageing	加令[齢], 老化	shear resistance	せん【剪】断
sensible heat	顕 熱		抵抗
sensitivity, susceptibility	感受性, 感度	sheath, leaf sheath	葉しょう【鞘】
		sheep	めん【緬】羊, 羊
sensitivity to daylength, photosensitivity	感光性	shell	殻, シェル
		shell, pod	さや【莢】
sensitivity to temperature	感温性	shelled corn	トウモロコシ粒
		shelter, shade	ひ【庇】蔭
sensory test	官能試験	shelter trees	避難林
separation, habitat segregation	すみわけ	shelter woods	ひ【庇】陰林
		shifting cultivation	焼畑農耕
septum, nuclear membrane	核 膜	shoot	若枝, 苗条, 茎(部)

shoot apex	茎頂	silage quality	サイレージ品質
shoot growth	茎葉［地上部，茎の］成［生］長	silica	シリカ，無水ケイ酸
		silicic acid	ケイ酸
short duration ［short time rotation, rapid rotation］grazing	短期輪換放牧	silicification	けい【珪】化
		silicon, Si	ケイ素
		silking stage	絹糸抽出期
short grass	短茎イネ科草，短草型草種	silo	サイロ
		silo ［silage］unloader	サイレージ取出し機
short grass pasture	短草型放牧草地		
short grass type	短草型	silt	微砂
short time rotation ［short duration, rapid rotation］grazing	短期輪換放牧	silvergrass-type ［Miscanthus-type］grassland	ススキ型草地
		simple correlation	単（純）相関
		simple improvement of grassland	草地簡易改良
short-day plant	短日植物		
short-day treatment	短日処理	simple leaf	単葉
short-lived grass	短年生草	simple pasture establishment	簡易草地造成
shrub, scrub	低かん【灌】木		
shuttle vector	シャトルベクター	simple ［rough］renovation	簡易更新
sick soil, soil sickness	いや【忌】地		
side manuring	側方施肥	simplex genotype	単式遺伝子型
sidework farmer, part-time farm household	兼業農家	simulation model	シミュレーションモデル
		single cell protein, SCP	単細胞タンパク【蛋白】質
sieve tube	師［篩］管		
sigmoid growth curve	S字型成［生］長曲線	single cropping, one crop system	一毛作
sign	標徴，徴候	single cropping	単作
significance test	有意性検定	single cross	単交配
significant difference	有意差	single leaf	個葉
significant digit	有効数字	single ［spaced］planting	1本植，個体植
significant level, level of significance	有意水準		
		single-crop seeding ［sowing］	単播
silage	サイレージ		
silage ［silo］unloader	サイレージ取出し機	site, location	立地
		size distribution	サイズ分布
silage additive	サイレージ添加物	size of farm	経営規模
		skin temperature	皮膚温
silage fermentation	サイレージ発酵	slag	スラグ，鉱さい
silage making, ensiling	サイレージ調製	slaked lime	消石灰

slant [slope] angle, inclination	傾斜（角）度	sod formation	草床形成
		sod grass	草床をつくる草
slippage	滑り率	sod-culture	草生栽培
slope	斜面，傾斜	sodium chloride, salt	食塩，塩化ナトリウム
slope, slope land, sloping land	傾斜地	sodium hydroxide	水酸化ナトリウム
slope, slope face	法　面		
slope [slant] angle, inclination	傾斜（角）度	sodium, Na	ナトリウム
		sod-seeding [sowing]	不耕起播き
slope aspect	傾斜［斜面］方位	soil	土　壌
		soil acidity	土壌酸性
slope face, slope	法　面	soil aggregate	土壌団粒
slope land, sloping land, slope	傾斜地	soil amelioration [amendment, improvement]	土壌改良
sloping land, slope land, slope	傾斜地	soil amendment [amelioration, improvement]	土壌改良
sloping [hillside] meadow	傾斜採草地		
sloping pasture, hillside grassland	傾斜草地	soil amendment matter, soil conditioner	土壌改良剤
slow [delayed] release fertilizer	緩効性［遅効性］肥料	soil animal	土壌動物
sludge	汚泥，スラッジ	soil chemistry	土壌化学
slurry	スラリー，泥状物	soil classification	土壌分類
		soil colloid	土壌コロイド
slurry barnyard manure	液状きゅう【厩】肥	soil conditioner, soil amendment matter	土壌改良剤
slurry injector	スラリーインジェ［ゼ］クタ	soil conservation	土壌保全
		soil crushing, clod breaking	砕　土
slurry pump	スラリーポンプ		
slurry separator	固液分離機	soil degradation	土壌退化
slurry spreader	スラリー散布機	soil depth	土壌深度
small implement	小農具	soil diagnosis	土壌診断
smoke injury	煙　害	soil disinfection [sterilization]	土壌殺菌
snow damage	雪　害		
snow endurance	耐雪性	soil dressing	客　土
snow zone [area, region]	積雪地帯	soil drying effect	乾土効果
		soil erosion	土壌侵食
soaking	浸　漬	soil fauna	土壌動物相
sociability	群　度	soil fertility	土壌肥よく【沃】度，地力
social insect	社会性昆虫		
sociation	基群集		

soil fumigation	土壌くん【燻】蒸	soil-borne disease	土壌伝染病
soil horizon	土壌層位	soil-exhaustive plant	地力消耗植物
soil improvement [amendment, amelioration]	土壌改良	soiling, fresh forage feeding, zero grazing	青刈り, 生草［青刈］給与
soil inoculation	土壌接種	soiling forage crop	青刈飼料作物
soil map	土壌図	solar energy	太陽エネルギー
soil microbiology	土壌微生物学	solar radiation, amount of solar radiation	日射（量）
soil microorganism	土壌微生物		
soil moisture	土壌水分	solid corrected milk, SCM	固形分補正乳
soil moisture [water] content	土壌水分含量	solid phase	固　相
soil organic matter	土壌有機物	solids-not-fat, SNF	無脂固形物
soil organism	土壌生物	soluble protein, CPs	溶解性タンパク【蛋白】質
soil packing, compacting, firming	鎮　圧	solution [water] culture, hydroponics	水　耕
soil pest	土壌害虫類	somaclonal variation	ソマクローナル［体細胞］変異
soil pH	土壌 pH		
soil physical property	土壌物理性		
soil physics	土壌物理学		
soil pollution	土壌汚染	somatic embryogenesis	体細胞はい【胚】形成
soil [land] preparation, leveling of ground	整　地		
		somatic hybrid	体細胞雑種
soil productivity	土壌生産力	source of variation	変動因
soil profile	土壌断面	source-sink relation	ソース・シンク関係
soil property	土　質		
soil reaction	土壌反応	south-western warm region	西南暖地
soil sampler	採土器		
soil science, pedology	土壌学	sowing, seeding	播種, たねまき
soil series	土壌統	sowing [seed, seeding] bed	播種床
soil sickness, sick soil	いや【忌】地		
soil sterilization [disinfection]	土壌殺菌	sowing [seeding] date	播種日
		sowing [seeding] depth	播種深（度）
soil structure	土壌構造		
soil surface	地　表	sowing [seeding] machine, seeder	播種機
soil survey	土壌調査		
soil temperature	地　温	sowing [seeding] method	播種法
soil texture	土　性		
soil type	土壌型	sowing [seeding] time	播種（時）期
soil water	土壌水	sown [artificial] grassland	人工草地
soilage, green fodder	青刈飼料		

sown [tame, artificial] pasture	牧草地	[dressing] split plot experiment	分割区試験法
soybean meal	大豆かす【粕】	spontaneous, native, indigenous	自生（の）, 在来（の）
spaced [single] planting	個体植, 1本植	spore	胞子, 芽胞
spaced planting, hill seeding [sowing]	点播	spore-forming bacteria	有胞子細菌
		sporulation	胞子形成
		spot [patch] grazing	パッチ状採食
spacer	スペーサ	spot treatment	局地処理
spacing	栽植距離	spray inoculation	噴霧接種
sparse planting	疎植	sprayer	噴霧機, 散布器
sparse seeding [sowing]	薄播き	spreader	展着剤
		spreader	延展器
spatial distribution	空間分布	spreading type	ひろがり型
spatial heterogeneity	空間的異質性, 空間不均一性	spring dressing	春肥
		spring habit	春播性
spatial variation	空間変異	spring seeding [sowing]	春播き
spatiotemporal	時空的な		
specialized farming	単一経営	sprinkler irrigation	スプリンクラーかんがい【灌漑】
species, race	種		
species diversity	種多様性		
species richness	種の豊富さ	sprout	ほう【萌】芽
species-area curve	種数面積曲線	sprout [coppice] forest	ほう【萌】芽林《林業》
specific activity	比活性, 比放射能		
		sprouting on the panicle, vivipary	穂発芽
specific combining ability	特定組合せ能力		
		square harrow	方形ハロー
specific conductivity	比伝導度	squash method	押しつぶし法
specific gravity	真比重, 比重	stable, barn	畜舎
specific leaf area, SLA	比葉面積	stable [barnyard] manure, manure	畜肥, きゅう【厩】肥, 肥【こえ】
specific pathogen free, SPF	寡菌動物		
spectrophotometer	分光光度計	stable isotope	安定同位体[元素]
spectrophotometric analysis	分光分析		
		stack	野積み, スタック
spherical virus	球形ウイルス		
spike	花穂, 穂, 穂状花序	stack silo	スタックサイロ
		stage of succession	遷移段階
spike tooth harrow	歯かんハロー	stalk, stem, haulm	茎
spikelet	小穂	stalk, culm	かん【稈】
spindly growth	徒長	stall	ストール, 牛床
split application	分施	stamen	雄しべ, 雄ずい

staminate [male] flower	雄花
stanchion	スタンチョン
stand	林分《個々の群落の》
stand	スタンド
stand, plant	株
stand preservation, foggage conservation	立毛貯蔵
standard	基準
standard deviation	標準偏差
standard error	標準誤差
standard variety	標準品種
standing	ちょ【佇】立
standing	静置
standing crop, biomass	現存量
standing dead (material), damping-off	立枯れ
starch	デンプン【澱粉】
starch granule	デンプン【澱粉】粒
starch value, SV	デンプン【澱粉】価
stationary drier	定置式乾燥装置
stationary hay baler, hay press	乾草圧縮機, ヘイプレス
statistic, statistical	統計学的
statistical, statistic	統計学的
statistical analysis	統計分析
statistical genetics	統計遺伝学
statistical survey	統計調査
statistical treatment	統計処理
statistics	統計学
steady density	安定密度
steaming	蒸煮
steep slope, steep sloping land	急傾斜地
steep sloping field	急傾斜畑
steep sloping land,	急傾斜地
steep slope	
steep sloping pasture	急傾斜草地
steer	去勢牛
stele	中心柱
stem, stalk, haulm	茎
stem bearing a flower bud, stem with a flower bud, flower stalk	着らい【蕾】枝
stem cutting	茎挿し
stem [tiller] weight	茎重
stem with a flower bud, flower stalk, stem bearing a flower bud	着らい【蕾】枝
stem/leaf ratio	茎葉比
steppe	ステップ《中央アジアの内陸草地》
sterile	不ねん【稔】(の)
sterile [aseptic] culture	無菌培養
sterile seed	不ねん【稔】種子
sterility, infertility	不ねん【稔】性
sterilization	滅[殺]菌, 滅菌法, 無菌化
steroid	ステロイド
stipule	たく【托】葉
stock [feeder] cattle	肥育用素牛
stock [stubble] density	株密度
stock [original] seed	原種《作物》, もと種
stock seed field	原種ほ【圃】
stock yard, paddock, yard	追込場
stock-holding agriculture	有畜農業
stocking, grazing	放牧
stocking [carrying, grazing] capacity	牧養力

stocking density, grazing density	放牧密度	structural carbohydrates	構造性炭水化物
stocking rate	単位面積当り放牧頭数	structure	構造
		structure of feed utilization	飼料構造
stolon, creeper, runner	ほふく【匍匐】茎	struggle for existence	生存競争
		stubble	刈株
stoloniferous, prostrate, creeping	ほふく【匍匐】型（の）	stubble grazing, stub-pasturage	刈（株）跡（地）放牧
stomach	胃	stub-pasturage, stubble grazing	刈（株）跡（地）放牧
stoma [pl. stomata]	気孔		
stomatal aperture	気孔開度	stump	根株《樹の》, スタンプ
stomatal infection	気孔感染		
stomatal resistance	気孔抵抗		
stomatal transpiration	気孔蒸散	stump cutter	スタンプカッタ
storability	貯蔵性	stump puller	抜根機
storage, preservation, conservation	貯蔵，保存	stunt	すくみ, 発育阻止, わい【矮】化
storage feeding	貯蔵草給与		
storage [reserve] organ	貯蔵器官	subalpine zone	亜高山帯
		subarctic zone	亜寒帯
storage root	貯蔵根	subclimax	亜極相，サブクライマックス
storage [reserve] tissue	貯蔵組織		
		subcrown internode	地中茎
store [fattening, feeding] cattle	肥育牛	subfamily	亜科
		sub-irrigation	地下かんがい【灌漑】
stover	わら《トウモロコシなどの》		
		submergence, flooding, waterlogging	湛水, 洪水
strain	系統		
stratification	階層構造, 層別化		
		submergence, overhead flooding	冠水
stratified clip method	層別刈取法		
stratum	階層, 層	subordinate species	従属種
straw	わら【藁】	sub-population	分集団
stress	ストレス	subsidy	補助金
string method	線状頻度法	subsoil	下層土, 心土
strip grazing	ストリップ放牧	subsoil improvement	土層【心土】改良
stripe [alley] cropping	帯状栽培		
stripe [row] seeding [sowing]	条播	subsoil plowing	心土耕
		subsoiling	心土破砕
		subspecies	亜種
stroma [pl. stromata]	ストロマ《クロロプラストの》	substance, material, matter	物質

substitution rate	代替比［率］	field-cured hay	
substrate, ground substance, matrix	基質	sun-leaf	陽葉
		sunshine recorder	日照計
subterranean [belowground, underground] part	地下部	superphosphate	過リン酸石灰
		supplement, supplementary [supplemental] feed	補助飼料
subterranean stem, rhizome	地下茎，根茎		
		supplemental [supplementary] feed, supplement	補助飼料
subtropical zone, subtropics	亜熱帯		
subtropics, subtropical zone	亜熱帯	supplemental [supplementary] feeding	飼料補給，補助飼料給与
suburban agriculture	近郊農業		
succeeding crop	後作	supplementary [supplemental] feed, supplement	補助飼料
succession	サクセッション，遷移		
successional equilibrium	遷移平衡	supplementary [supplemental] feeding	飼料補給，補助飼料給与
sucker	吸収器官		
sucker	台芽，ひこばえ	surface (layer) application	表層施肥
sucking insect	吸汁昆虫		
sucking [absorbing] root	吸収根	surface area	表面積
		surface horizon	表層
suckler cow	子付（繁殖）雌牛，授乳牛	surface irrigation	地表かんがい【灌漑】
suckling, nursing	ほ【哺】乳	surface run-off	表面［地表］流出［流去］
sucrose, saccharose	ショ糖，スクロース		
		surface sowing	表面播種
sugar	糖質，糖	survival, survivorship	生存
sulfur amino acid	含硫アミノ酸	survival [survivorship] curve	生存曲線
sulfur, S	イオウ		
summed dominance ratio, SDR	積算優占度	survival ratio	生存率
		surviving plant [ramet]	生存株
summer (season) grazing	夏季放牧		
		survivorship, survival	生存
summer cropping	夏作	survivorship [survival] curve	生存曲線
summer depression	夏枯れ		
summer forage crop	夏作飼料作物	susceptibility, sensitivity	感受性，感度
summer survival	越夏性		
sun plant	陽光［陽生］植物	susceptible	り【罹】病性の
		sustainable development	持続可能な開発
sun-cured hay,	自然乾草		

swallowing	えん【嚥】下
sward, grassland	草地
sward [plant] height	草高
swath, gathering width	刈幅
swath turner	反草機
sweep [buck] rake	スィープレーキ
swelling	膨潤
switch-back feeding experiment	反転飼養試験法
symbiosis, mutualism	共生，相利共生
symbiotic nitrogen fixation	共生的窒素固定
symptomless infection	無病徴感染
symptom	病徴
synecology	群（落）生態学
synthesis	合成
synthetic milk	人工乳
synthetic variety	合成品種
system	体系
system model	システムモデル
systematic arrangement	系統的配置
systematic sampling	系統的抽出
systematized techniques	技術体系
systemic infection	全身感染
systemic symptom	全身病徴
systems analysis	システム分析
systems approach	システムズアプローチ

T

tall-grass type	長草型
tame [sown, artificial] pasture	牧草地
tannic acid	タンニン酸
tannin	タンニン
tap [main] root	主根
tap [axial] root	直根
tassel	雄穂《トウモロコシの》
taxonomy	分類学
TDN content	TDN含量
TDN yield	TDN収量
teat	乳頭
tedder and rake	テッダレーキ
tedding	転草
telemeter	テレメータ
teleutospore	冬胞子
temperate	寒地型（の）
temperate [cool season] grass	寒地型イネ科牧草
temperate zone	温帯
temperature	温度
temperature response	温度反応
temporary grass	短年草
temporary grassland	一時的草地
temporary pasture	一時的放牧地
terminal	頂生（の）
terminal [apical] bud	頂芽
ternate compound leaf	三出複葉
terrace field	段畑
terraced paddy field, rice terrace	棚田
territory	縄張り，領域，テリトリ
Tertiary	第三紀層
tertiary wall	三次膜《細胞膜の》
test	検定
test field	試験ほ【圃】
test for resistance	抵抗性検定
test of specific character	特性検定
tetany	テタニー
tether grazing, tethering	繋牧，繋飼い
tethering, tether grazing	繋牧，繋飼い
tetraploid	四倍体（の）
tetrasomic inheritance	四染色体遺伝

tetravalent [quadrivalent] chromosome	四価染色体	harrowing, cultivation	
		tiller	耕うん【耘】機
thermal [heat] capacity	熱量	tiller	分げつ
		tiller bud	分げつ芽
thermal death point	致死温度	tiller composition	分げつ構成
thermal radiation	熱放射	tiller density	分げつ密度
thermocouple	熱電対	tiller number	分げつ数
thermograph, recording thermometer	自記温度計	tiller [stem] weight	茎重
		tillering habit	分げつ習性
		tillering node	分げつ節
thermometer	温度計	tillering stage	分げつ期
thermoperiodicity	感温周期性	tilling [plowing] method	耕起法
thiamin, vitamin B_1	チアミン, ビタミンB_1		
		time of fertilizer application	施肥時期
thick stand	密生（草生）		
thickening	肥大	tissue	組織
thickening growth	肥大成[生]長	tissue culture	組織培養
thinlayer chromatography, TLC	薄層クロマトグラフィ	titratable acidity	滴定酸度
		tocopherol, vitamin E	トコフェロール, ビタミンE
thinning	間引き		
thinning frequency	間伐頻度《林業》	tofu cake	豆腐かす【粕】
		tolerance, resistance	抵抗性, 耐性, 抵抗力
thinning intensity	間伐強度		
thousand kernel [seed] weight	千粒重	toluene distillation method	トルエン蒸溜法
three major nutrients, three primary nutrients, N,P,K elements	肥料三要素	top [aboveground, aerial] part	地上部
		top grass	上繁草
		top grazing	上層採食, トップグレージング
three phases of soil	土壌三相		
three primary nutrients, three major nutrients, N,P,K elements	肥料三要素		
		top unloader	トップアンローダ
		top-cross	トップ交雑
three-way cross	三系交配	topdressing, additional manure	追肥
threshing	脱穀		
thylakoid	チラコイド	topography	地形
thyroid gland	甲状腺	top-root ratio	T-R比
tick fever	ダニ熱	topsoil	表土
tillage, plowing and	耕うん【耘】	total carbon	全炭素

total digestible nutrients, TDN	可消化養分総量		《農業類型》
total mixed ration, TMR	混合飼料	transition matrix	推移行列
		translocation	転流，転座，移行
total nitrogen	全窒素	transmission	伝達，伝搬
total short wave [global, total solar] radiation	全天日射，全短波放射	transmission, infection	伝　染
		transmittance	透　過
		transovarial passage	経卵伝染
total solar [global, total short wave] radiation	全天日射，全短波放射	transpiration	蒸　散
		transpiration rate	蒸散速度
		transplant experiment	移植実験［試験］
total solid body	全固形物		
total sugar	全　糖	transplanting	移　植
tourism	観　光	transplanting culture	移植栽培
tower silo	タワーサイロ，塔サイロ	transport	輸　送
		tread power	踏圧力
toxicant	毒　物	treading, trampling, compacting	てい【蹄】傷，踏圧，踏付け
toxicity, virulence	毒　性		
toxin	毒　素	treatment	処理，治療
trace element, microelement, micronutrient, mior element	微量元素［要素］	tree for green manure and fodder	飼肥料木
		tree plantation, afforested land	植林地，造林地
trace mineral	微量ミネラル	tree stratum	高木層
tracer technique	トレーサー法	trench silo	トレンチサイロ
trachea	気管《動物》	trend	動向，傾向
tracheid	仮道管	tricarboxylic acid [TCA] cycle	トリカルボン酸回路，クエン酸回路
tractor	トラクタ		
trade-off	トレードオフ		
traditional agriculture [farming]	伝統(的)農法	trimming	掃除刈り
		tripping	トリッピング
trailer	トレーラ	tropical	熱帯の，暖地型(の)
trampling, treading, compacting	てい【蹄】傷，踏圧，踏付け		
		tropical grassland	熱帯草地
trampling [treading] damage	てい【蹄】傷害	tropical zone, tropics	熱　帯
		trough	飼　槽
trampling tolerance [resistance]	踏圧耐性，耐踏圧性	true protein	純タンパク【蛋白】質
transformation	形質転換	tuber	塊　茎
transgenic crop	遺伝子組換え作物	tubulous flower	管状花
		tuff	凝灰岩
transhumance	移牧	tufted, bunch, tussock	株型の

tufted [bunch] type	そう【叢】状型，そう【叢】生型，株型	underground creeper	地下ほふく【匍匐】枝
		underground drain	地下排水
tundra	ツンドラ	underground [subterranean, belowground] part	地下部
turbidity	濁度		
turf, lawn	芝(地)，芝生		
turnover	回転，ターンオーバー	underground silo	地下サイロ
		underground water	地下水
turning	反転	undergrowth, bottom grass, understory	下草，下繁草，下層植生
turning out to pasture, setting up the grazing	入牧		
		undernutrition	栄養不足
tussock, bunch, tufted	株型の	understory, bottom grass, undergrowth	下草，下繁草，下層植生
twin plant	双はい【胚】植物		
two season grazing	二シーズン放牧	unevenness of application	散布むら
two-way plow	互用プラウ		
typhoon	台風	ungulate	有てい【蹄】動物

U

		unhulled rice	もみ【籾】
		uniform distribution	一様分布
udder	乳房	unloader	アンローダ
ultracentrifuge	超遠心(分離)機	unparatable [wolf] grass [plant]	不食草
ultrafiltration	限外濾過	unplowed grssland establishment, non-tilled establishment	不耕起草地造成
ultrastructure	微細構造		
ultrathin section	超薄切片		
ultraviolet absorption method	紫外吸光度法		
		unplowed reseeding	不耕起追播
ultraviolet radiation	紫外線照射	unsaturated fatty acid	不飽和脂肪酸
ultraviolet ray	紫外線	upland [field] crop	畑作物
umbel	散形花序	upland field in paddy-upland rotation	輪換畑
unavailable [non-available] element	不可給態要素		
		upland grassland	高地草地
uncultivated [abandoned] land	放棄地	upland meadow	山地草原
		upland soil	台地[畑]土壌
undegradable protein	非分解性タンパク【蛋白】質	upright habit	直立性
		uptake, absorption	吸収
underdrainage	暗きょ【渠】排水	urea	尿素
		urease	ウレアーゼ
underfeeding	低栄養飼養	urediniospore	夏胞子

urination	排尿
urine	尿
urine spreader	尿散布機
urine scorch	尿やけ
use efficiency, efficiency of utilization	利用（効）率
useful life, durable years	耐用年数
useful wild herb	有用野草

V

vacuole	液胞
vacuum dryer	真空乾燥機
vacuum pump	真空ポンプ
vacuum silo	真空サイロ
variability	変異性，可変性
variance	分散
variance ratio	分散比
variate	変量
variation	変異
variation	変動，異形
varietal certification	品種証明
varietal characteristics	品種特性
variety	品種
variety preservation	品種保存
variety test	品種比較試験
vascular bundle	維管束
vascular bundle sheath	維管束しょう【鞘】
vascular disease	維管束病
vase, calyx	がく
veal calf, calf	子牛
vector	病害ばい【媒】介者
vector	ベクター
vegetable	野菜，そ【蔬】菜
vegetation	植生
vegetation dynamics	植生動態
vegetation in forest	林床植生
understory, ground vegetation	
vegetation map	植生図
vegetation pattern	植生パターン
vegetation structure	植生構造
vegetational continuum	植生連続
vegetational replacement	植生交替
vegetational transition	植生推移
vegetative growth	栄養成［生］長
vegetative organ	栄養器官
vegetative propagation	挿木［芽］
vegetative reproduction	栄養（体）生殖［繁殖］
vegetative stage	栄養成［生］長期
vein	脈，静脈
vein banding	葉脈緑帯
vermiculite	バーミキュライト
vernalization	春化，バーナリゼーション
vertical distribution	垂直分布
vesicular arbuscular mycorrhizas	のう状樹枝状菌根，VA菌根
vessel	道管《植物》
viability of seed, germination ability	発芽勢
viable farm	自立経営
vigor	活力
vine	つる
viny	つる性（の）
virescence	異常緑変
virgin forest	原生林
virgin soil	未耕土
virulence, pathogenicity	病原力［性］
virulence, toxicity	毒性
virus	ウイルス
virus disease	ウイルス病
viscosity	粘度，粘性

visible ray	可視光線	weigher	
vitality	活力度	walking speed	歩行速度
vitamin	ビタミン	walking tractor	歩行トラクタ
vitamin A_1, retinol	ビタミン A_1, レチノール	warm region	暖　地
		warm temperate zone	暖温帯
vitamin B_1, thiamin	ビタミン B_1, チアミン	warmth index	暖かさの指数
		warm-up grazing	じゅん【馴】致放牧
vitamin B_2, riboflavin	ビタミン B_2, リボフラビン	waste	廃棄物, ふん【糞】尿
vitamin B_6, pyridoxin	ビタミン B_6, ピリドキシン	waste [degrading, deteriolated] grassland	衰退［荒廃］草地
vitamin C, ascorbic acid	ビタミン C, アスコルビン酸		
		waste water	廃　水
vitamin D, calciferol	ビタミン D, カルシフェロール	water absorption	吸　水
		water balance	水分平衡
		water budget [balance]	水収支
vitamin E, tocopherol	ビタミン E, トコフェロール	water [moisture] capacity	容水量
		water conservation	水保全, 水のかん【涵】養
vivipary, sprouting on the panicle	穂発芽		
		water [moisture] content	含水量, 水分含量
volatile basic nitrogen, VBN	揮発性塩基態窒素		
		water [solution] culture, hydroponics	水　耕
volatile constituent	揮発性成分		
volatile fatty acid, VFA	揮発性脂肪酸		
		water erosion	水　食
volatile substance [matter]	揮発性物質	water extract	水（溶性）抽出物
volatilization	揮　発	water holding capacity	保水力［性］
volcanic ash	火山灰	water management	水管理
volcanic ash soil	火山灰土壌	water pollution	水質汚濁［汚染］
voluntary intake	自由採食（量）		
volunteering, natural reseeding	自然下種	water potential	水ポテンシャル
		water quality	水　質
		water requirement	要水量
W		water resisting property	耐水性
waferer	ウエファー成形機	water soluble carbohydrate, WSC	水溶性糖類［炭水化物］
wagon	ワゴン	water soluble phosphate	水溶性リン酸
walk throw type	歩行型体重計		

water soluble protein	水溶性タンパク【蛋白】質	whole crop	ホールクロップ
water soluble vitamin	水溶性ビタミン	whole milk	全乳
water stress	水ストレス	whole-crop silage	ホールクロップサイレージ
water stress resistance	水ストレス耐性	whole-day grazing	昼夜放牧
water uptake	水吸収	wild animal	野生動物
water use	水利用	wild flower	ワイルドフラワー
water use efficiency	水利用効率		
watering	かん【灌】水	wild grassland used for grazing, native pasture	野草放牧地
watering place	給水場		
watering point, guzzler, drinker	飲水施設		
		wild plant	野生植物, 野草
waterlogging, flooding, submergence	湛水, 洪水	wild race	野生種
		wild relative	近縁野生種
		wildlife	野生生物
watershed	集水域	wildlife conservation	野生生物保全【保護】
water-soaked	水浸状の		
water-stable aggregate	耐水性団粒	wilting	しおれ
wavelength	波長	wilting, prewilting	予乾
wax	ワックス, ロウ	winch	ウインチ
waxy, glutinous	もち【糯】性の	wind	風
weak tiller	弱小分げつ	wind damage	風害
weaning	離乳	wind dispersal	風散布
weather	天気	wind erosion	風食
weathering	風化	wind hedge	防風垣
weed	雑草	wind selection, winnowing	風選
weed control	雑草防除		
weeder	除草機	wind velocity	風速
weeding	除草	wind-break forest	防風林
weight	重量	windrow	集草列, ウインドロー, 地干し列
weight ratio	重量比		
wet endurance	耐湿性		
wetland	湿地	winnowing, wind selection	風選
wettable powder	水和剤		
wheat bran	ふすま《表皮》	winter (season) grazing	冬季放牧
wheat straw	麦わら【藁】		
wheel cutter	ホイールカッタ	winter annual	越年草
wheel track [rut]	輪だち【轍】, 装輪トラック	winter bud	冬芽
		winter cereal	冬穀物
		winter [secondary] cropping	裏作
wheel tractor	車輪型トラクタ		
whisky by-product feed	ウィスキーかす【粕】	winter cropping	冬作

winter cropping on drained paddy field	水田裏作
winter habit	秋播性
winter hardiness	耐冬性
winter housing system	夏山冬里方式
winter killing	冬枯れ
winter survival	越冬性
wintering, overwintering	越冬
wintering, hibernation	冬眠
wolf [unparatable] grass [plant]	不食草
woodland, forestland	樹林地，林地，森林
woody [arboreous] plant, arbor	木本類
wool production	羊毛生産
work [operation] system	作業体系
workable days	可働日数
working efficiency	作業効率
working machine	作業機
wrapped silage	ラップサイレージ
wrapping	ラッピング

X

xerophyte	乾生植物
xylem	木部
xylem parenchyma	木部柔組織
xylose	キシロース

Y

yard, paddock, stock yard	追込場
yearlong grazing	周年放牧
yearlong outdoor feeding	屋外周年飼養
yearly [year-to-year] correlation	年次相関
year-round	周年（の）
year-round culture	周年栽培
year-to-year [yearly] correlation	年次相関
yeast	酵母，イースト
yellow-ripe stage	黄熟期
yield	収量，収率
yield component	収量構成要素
yield curve	収量曲線
yield decrease	減収
yield increase	増収
young tree	若木
young ear [panicle] formation stage	幼穂形成期
youngest expanding leaf	最上位展開葉

Z

zeolite	ふつ【沸】石，ゼオライト
zero grazing, soiling, fresh forage feeding	青刈り，生草[青刈]給与
zerotillage, non-tilled cropping, plowless farming	不耕起栽培
zigzag planting	千鳥植
zinc, Zn	亜鉛
zonation	帯状[成帯]分布，ゾーネーション
zone	地帯
zone electrophoresis	ゾーン電気泳動，分域電気泳動
zoosporangium [pl. zoosporangia]	遊走子のう
zoospore	遊走子
zootechnical [animal] science, zootechnology	畜産学

zootechnology, animal [zootechnical] science	畜産学	Zoysia pasture, Zoysia-type grassland, Zoysia dominated grass	シバ(型)草地
Zoysia dominated grassland, Zoysia-type grassland, Zoysia pasture	シバ(型)草地	Zoysia-type grassland, Zoysia dominated grassland, Zoysia pasture	シバ(型)草地

III 植物名・病害名・昆虫および動物名

植物名（和名順）……………………………173
植物名（学名順）……………………………197
病害名（和名順）……………………………219
病害名（学名順）……………………………225
昆虫および動物名（和名順）………………231
昆虫および動物名（学名順）………………241

植 物 名 （和名順）

和 名	学 名	英 名
アオカモジグサ	*Agropyron racemiferum* (Steud.) Koidz.	wheatgrass
アオゲイトウ	*Amaranthus retroflexus* L.	green amaranth
アオツヅラフジ	*Cocculus orbiculatus* (L.) Forman	coralbeads
アオビユ	*Amaranthus viridis* L.	slender amaranth
アオヤギソウ	*Veratrum maackii* Regel var. *parviflorum* (Miq.) Hara et Mizushima	false hellebore
アカクローバ, アカツメクサ	*Trifolium pratense* L.	red clover
アカザ	*Chenopodium album* L. var. *centrorubrum* Makino	goose-foot pigweed
アカソ	*Boehmeria tricuspis* (Hance) Makino	false nettle
アカツメクサ, アカクローバ	*Trifolium pratense* L.	red clover
アカネ	*Rubia akane* Nakai	madder
アカマツ	*Pinus densiflora* Sieb. et Zucc.	Japanese red pine
アキカラマツ	*Thalictrum minus* L. var. *hypoleucum* (Sied. et Zucc.) Miq.	meadow-rue
アキグミ	*Elaeagnus umbellata* Thunb.	oleaster
アキノエノコログサ	*Setaria faberi* Herrm.	foxtail grass, bristle grass
アキノキリンソウ	*Solidago virgaurea* L. var. *asiatica* Nakai	goldenrod
アキノノゲシ	*Lactuca indica* L. var. *laciniata* (O. Kuntze) Hara	milkweed
アキメヒシバ	*Digitaria violascens* Link	small crabgrass, violet crabgrass
アズマギク	*Erigeron thunbergii* A. Gray	
アズマネザサ	*Pleioblastus chino* (Franch. et Savat.) Makino	
アゼガヤツリ	*Cyperus globosus* All.	
アゼテンツキ	*Fimbristylis squarrosa* Vahl	
アセビ	*Pieris japonica* (Thunb.) D. Don	lily-of-the valley bush

植物名（和名順）

アブラススキ	*Eccoilopus cotulifer* (Thunb.) A. Camus	
アメリカアサガオ	*Ipomoea hederacea* (L.) Jacq.	
アメリカイヌホオズキ	*Solanum americanum* Mill.	
アメリカオニアザミ	*Cirsium vulgare* (Savi) Tenore	bull thistle
アメリカキンゴジカ	*Sida spinosa* L.	
アメリカセンダングサ	*Bidens frondosa* L.	devils beggar-ticks, beggar-ticks, stick-tight
アメリカツノクサネム	*Sesbania exaltata* Cory et Rydb.	
アメリカネナシカズラ	*Cuscuta pentagona* Engelm.	
アメリカホオズキ	*Physalis heterophylla* Nees	
アリスクローバ	*Alysicarpus vaginalis* (L.) DC.	alyce clover
アリノトウグサ	*Haloragis micrantha* (Thunb.) R. Br.	creeping raspwort
アルサイククローバ	*Trifolium hybridum* L.	alsike clover
アルファルファ, ルーサン, ムラサキウマゴヤシ	*Medicago sativa* L.	alfalfa, lucerne
アレチウリ	*Sicyos angulatus* L.	
アレチギシギシ	*Rumex conglomeratus* Murr.	cluster dock
アレチノギク	*Erigeron bonariensis* L.	
アレチノチャヒキ	*Bromus sterilis* L.	
アレチマツヨイグサ	*Oenothera parviflora* L.	evening primrose
アワ	*Setaria italica* Beauv.	foxtail millet, barngrass, Italian millet, Chinese corn
イ	*Juncus effusus* L. var. *decipiens* Buchen.	soft rush, mat rush, rush
イタドリ	*Polygonum cuspidatum* Sieb. et Zucc.	Japanese knotweed
イタリアンライグラス, ネズミムギ	*Lolium multiflorum* Lam.	Italian ryegrass
イチゴツナギ	*Poa sphondylodes* Trin.	bluegrass, meadowgrass
イチビ	*Abutilon theophrasti* Medic.	velvetleaf, piemaker
イトコヌカグサ	*Agrostis capillaris* L.	
イトススキ	*Miscanthus sinensis* Anderss. f. *gracillimus* (Hitchc.) Ohwi	eulalia
イヌコリヤナギ	*Salix integra* Thunb.	willow

イヌタデ	*Polygonum longisetum* De Bruyn	knotweed, smartweed
イヌビエ	*Echinochloa crus-galli* (L.) Beauv.	barnyardgrass
イヌビユ	*Amaranthus lividus* L.	livid amaranth
イヌホオズキ	*Solanum nigrum* L.	hound-berry, fox-grape
イヌムギ, レスクグラス	*Bromus catharticus* Vahl	rescuegrass
イネ	*Oryza sativa* L.	rice
イワノガリヤス	*Calamagrostis langsdorffii* (Link) Trin	small-reed
インターミーディエトホィートグラス	*Agropyron intermedium* P. Beauv.	intermediate wheatgrass
ウィーピングラブグラス, シナダレスズメガヤ	*Eragrostis curvula* Nees	weeping lovegrass
ウイメラライグラス	*Lolium rigidum* Gaud.	wymmera ryegrass
ウェスタンホィートグラス	*Agropyron smithii* Rydb.	western wheatgrass
ウキアゼナ	*Bacopa rotundifolia* Wettst.	disc waterhyssop
ウサギアオイ	*Malva parviflora* L.	
ウシクサ	*Andropogon brevifolius* Swartz	beardgrass
ウシノケグサ, シープフェスク	*Festuca ovina* L.	sheep's fescue
ウシノシッペイ	*Hemarthria sibirica* (Gandog.) Ohwi	
ウシハコベ	*Stellaria aquatica* (L.) Scop.	giant chickweed
ウツボグサ	*Prunella vulgaris* L. var. *lilacina* (Nakai) Nakai	selfheal
ウド	*Aralia cordata* Thunb.	angelica tree
ウマゴヤシ, バークローバ	*Medicago polymorpha* L.	bur clover
ウマノアシガタ	*Ranunculus japonicus* Thunb.	buttercup, crowfoot
ウマノチャヒキ	*Bromus tectorum* L.	downy brome
ウメバチソウ	*Parnassia palustris* L.	grass-of-parnassus, white buttercups
ウリクサ	*Vandellia crustacea* (L.) Benth.	
エゾタンポポ	*Taraxacum hondoense* Nakai ex H. Koidz.	dandelion, blowballs
エゾニュウ	*Angelica ursina* (Rupr.) Maxim.	angelica
エゾノギシギシ	*Rumex obtusifolius* L.	broadleaf dock
エノコログサ	*Setaria viridis* (L.) Beauv.	green bristle-grass, green foxtail
エビスグサ	*Cassia obtusifolia* L.	sicklepod

植物名（和名順）

エンバク	*Avena sativa* L.	oat
オオアザミ	*Silybum marianum* Gaertn.	
オオアブラススキ	*Spodiopogon sibiricus* Trin.	
オオアレチノギク	*Erigeron sumatrensis* Retz.	
オオアワガエリ, チモシー	*Phleum pratense* L.	timothy
オオイタドリ	*Polygonum sachalinense* Fr. Schm.	sachaline giant knotweed
オオイチゴツナギ	*Poa nipponica* Koidz.	bluegrass, meadow-grass
オオイヌタデ	*Polygonum lapathifolium* L.	pale persicaria, willow smartweed, pale smartweed
オオイヌノフグリ	*Veronica persica* Poir.	birdseye, speedwell, cat's eyes
オオウシノケグサ, レッドフェスク, クリーピングフェスク	*Festuca rubra* L.	red fescue, creeping fescue
オオオナモミ	*Xanthium occidentale* Bertoloni	
オオカッコウアザミ	*Ageratum houstonianum* Mill.	
オオカニツリ, トールオートグラス	*Arrhenatherum elatius* (L.) Mertens et Koch	tall oatgrass, false oatgrass
オオクサキビ	*Panicum dichotomiflorum* Michx.	fall panicum
オオケタデ	*Polygonum pilosum* Roxb.	
オオシラタマソウ	*Silene conoidea* L.	
オオスズメノカタビラ	*Poa trivialis* L.	rough stalked meadowgrass
オオセンナリ	*Nicandra physaloides* Gaertn.	
オオヂシバリ	*Ixeris debilis* (Thunb.) A. Gray	
オオチドメ	*Hydrocotyle ramiflora* Maxim.	water-pennywort, navelwort
オーチャードグラス, カモガヤ	*Dactylis glomerata* L.	orchardgrass, cocksfoot
オオニシキソウ	*Euphorbia maculata* L.	
オオバコ	*Plantago asiatica* L.	plantain
オオハンゴンソウ	*Rudbeckia laciniata* L.	cut-leaf coneflower
オオブタクサ	*Ambrosia trifida* L.	giant ragweed
オオホナガアオゲイトウ	*Amaranthus palmeri* S. Wats.	
オオマツヨイグサ	*Oenothera erythrosepala* Borbas	evening primrose
オオムギ	*Hordeum vulgare* L.	barley
オオヤマフスマ	*Moehringia lateriflora* (L.) Fenzl	grove-sandwort
オガサワラスズメノヒエ	*Paspalum conjugatum* Bergius	

オカトラノオ	*Lysimachia clethroides* Duby	loosestrife
オガルカヤ	*Cymbopogon tortilis* (Presl) A. Camus var. *goeringii* (Steud.) Hand.-Mazz.	
オギ	*Miscanthus sacchariflorus* (Maxim.) Benth.	eulalia
オキナアサガオ	*Jacquemontia tamnifolia* (L.) Griseb.	
オキナグサ	*Pulsatilla cernua* (Thunb.) Spreng.	pasque-flower
オキナワミチシバ	*Chrysopogon aciculatus* (Rstz.) Trin.	
オトギリソウ	*Hypericum erectum* Thunb.	St. John's-wort
オトコエシ	*Patrinia villosa* (Thunb.) Juss.	
オトコヨモギ	*Artemisia japonica* Thunb.	wormwood
オナモミ	*Xanthium strumarium* L.	cocklebur
オニウシノケグサ, トールフェスク	*Festuca arundinacea* Schreb.	tall fescue
オニシバ	*Zoysia macrostachya* Franch. et Savat.	
オニタビラコ	*Youngia japonica* (L.) DC.	oriental hawkbeard
オニノゲシ	*Sonchus asper* (L.) Hill	spiny sowthistle
オヒゲシバ	*Chloris virgata* Swartz	feather top rhodesgrass, feather fingergrass
オヒシバ	*Eleusine indica* (L.) Gaertn.	goosegrass
オミナエシ	*Patrinia scabiosaefolia* Fisch.	
オランダミミナグサ	*Cerastium glomeratum* Thuill.	sticky mouse-ear, chickweed
カーペットグラス	*Axonopus affinis* Chase	carpetgrass
カウピー	*Vigna sinensis* (L.) Savi ex Hassk.	cowpea
カキネガラシ	*Sisymbrium officinale* Scop.	
ガクアジサイ	*Hydrangea macrophylla* (Thunb.) Ser. f. *normalis* (Wilson) Hara	hydrangea
カシワ	*Quercus dentata* Thunb.	daimyo oak
カスマグサ	*Vicia tetrasperma* (L.) Schreb.	lentillare, smooth tare
カゼクサ	*Eragrostis ferruginea* (Thunb.) Beauv.	lovegrass
カセンソウ	*Inula salicina* L. var. *asiatica* Kitam.	willow-leaved-inule

植物名（和名順）

カタバミ	*Oxalis corniculata* L.	creeping wood sorrel, creeping lady's-sorrel
カナダブルーグラス, コイチゴツナギ	*Poa compressa* L.	Canada bluegrass
カナビキソウ	*Thesium chinense* Turcz.	bastard-toad-flax
カニツリグサ	*Trisetum bifidum* (Thunb.) Ohwi	
カブラブラグラス	*Panicum coloratum* L. (Kabulabula type)	kabulabulagrass
ガマズミ	*Viburnum dilatatum* Thunb.	viburnum, guelder-rose
カミツレ	*Matricaria chamomilla* L.	
カミツレモドキ	*Anthemis cotula* L.	
カモガヤ, オーチャードグラス	*Dactylis glomerata* L.	orchardgrass, cocksfoot
カモジグサ	*Agropyron tsukushiense* (Honda) Ohwi var. *transiens*(Hack.) Ohwi	drooping wheatgrass
カヤツリグサ	*Cyperus microiria* Steud.	chufa, flatsedge, umbrella sedge
カラードギニアグラス	*Panicum coloratum* L.	colored guineagrass
カラクサガラシ, カラクサナズナ	*Coronopus didymus* Smith	
カラクサナズナ, カラクサガラシ	*Coronopus didymus* Smith	
カラスノエンドウ, コモンベッチ	*Vicia sepium* L.	common vetch
カラスムギ	*Avena fatua* L.	wild oat
カリヤス	*Miscanthus tinctorius* (Steud.) Hack.	eulalia
カリヤスモドキ	*Miscanthus oligostachyus* Stapf	eulalia
カワラケツメイ	*Cassia nomame* (Sieb.) Honda	senna
カワラサイコ	*Potentilla chinensis* Ser.	cinquefoil, potentil, five-finger
カワラナデシコ	*Dianthus superbus* L. var. *longicalycinus* (Maxim.) Williams	superb-pink
カンショ（甘藷）, サツマイモ	*Ipomoea batatas* (L.) Lam.	sweet potato
キキョウ	*Platycodon grandiflorum* (Jacq.) A.DC.	ballon flower, Chinese bellflower, Japanese bellflower

植物名（和名順）　[179]

キクユグラス	*Pennisetum clandestinum* Hochst. ex Chiov.	kikuyugrass
ギシギシ	*Rumex japonicus* Houtt.	dock
キジムシロ	*Potentilla fragarioides* L. var. *major* Maxim.	cinquefoil, potentil, five-finger
キシュウスズメノヒエ	*Paspalum distichum* L.	knotgrass, water couch
キゾメカミツレ	*Anthemis arvensis* L.	
キツネアザミ	*Hemistepta lyrata* Bunge	
キツネノボタン	*Ranunculus silerifolius* Lev.	buttercup, crowfoot
キツネヤナギ	*Salix vulpina* Anders.	willow
ギニアグラス	*Panicum maximum* Jacq.	guineagrass
キバナアルファルファ	*Medicago falcata* L.	yellow alfalfa
キバナスィートクローバ，シナガワハギ	*Melilotus officinalis* Pallas	
キハマスゲ，ショクヨウガヤツリ	*Cyperus esculentus* L.	yellow nutsedge
キビ	*Panicum miliaceum* L.	millet, common millet
ギョウギシバ，バーミューダグラス	*Cynodon dactylon* (L.) Pers.	bermudagrass
キランソウ	*Ajuga decumbens* Thunb.	bugle
キレハイヌガラシ	*Rorippa sylvestris* Bess.	
キンエノコロ	*Setaria glauca* (L.) Beauv.	yellow bristlegrass, yellow foxtailgrass, bottlegrass
キンギンナスビ	*Solanum aculeatissimum* Jacq.	
ギンネム	*Leucaena leucocephala* (Lam.) De Wit.	leucaena
キンミズヒキ	*Agrimonia japonica* (Miq.) Koidz.	agrimony, cocklebur
クサイ	*Juncus tenuis* Willd.	path rush
クサフジ	*Vicia cracca* L.	bird vetch, tufted vetch, gerard vetch
クサヨシ，リードカナリーグラス	*Phalaris arundinacea* L.	reed canarygrass
クズ	*Pueraria lobata* (Willd.) Ohwi	kudzu-vine, kudzu
クスダマツメクサ	*Trifolim campestre* Schreb.	
クヌギ	*Quercus acutissima* Carruth.	Japanese chestunt oak
クマイザサ	*Sasa senanensis* (Franch. et Savat.) Rehder	sasa, dwarf bamboo
クマイチゴ	*Rubus crataegifolius* Bunge	bramble
クマザサ	*Sasa veitchii* (Carr.) Rehder	sasa, dwarf bamboo

植物名（和名順）

和名	学名	英名
クララ	*Sophora flavescens* Solander ex Aiton var. angustifolia (Sieb. et Zucc.) Kitagawa	sophora
クリ	*Castanea crenata* Sieb. et Zucc.	chestnut
クリーピングフェスク, オオウシノケグサ, レッドフェスク	*Festuca rubra* L.	red fescue, creeping fescue
クリーピングベントグラス, ハイコヌカグサ	*Agrostis stolonifera* L.	creeping bentgrass
グリーンパニック	*Panicum maximum* Jacq. var. *trichoglume* Eyles	green panic
グリーンリーフデスモディウム	*Desmodium intortum* (Mill.) Urb.	greenleaf desmodium, kuru vine
クリノイガ	*Cenchrus brownii* Roem. et Schult.	
クリムソンクローバ	*Trifolium incarnatum* L.	crimson clover
クレステッドホィートグラス	*Agropyron cristatum* Gaertn.	crested wheatgrass
クロタラリア	*Crotalaria juncea* L.	sun hemp
クロマツ	*Pinus thunbergii* Parlat.	Japanese black pine
グンバイナズナ	*Thlaspi arvense* L.	
ケチヂミザサ	*Oplismenus undulatifolius* (Ard.) Roemer et Schultes	
ゲンゲ, レンゲ	*Astragalus sinicus* L.	Chinese milk vetch
ケンタッキーブルーグラス, ナガハグサ	*Poa pratensis* L.	Kentucky bluegrass
ゲンノショウコ	*Geranium thunbergii* Sieb. et Zucc.	carnesbill
コアカザ	*Chenopodium ficifolium* Smith	fig-leaved goose foot
コイチゴツナギ, カナダブルーグラス	*Poa compressa* L.	Canada bluegrass
ゴウシュウアリタソウ	*Chenopodium carinatum* R. Br.	
コウシュンシバ	*Zoysia matrella* (L.) Merr.	Manilagrass, Manila-bluegrass
コウボウ	*Hierochloe bungeana* Trin.	
コウボウムギ	*Carex kobomugi* Ohwi	sedge
コウマゴヤシ	*Medicago minima* (L.) Bartal.	little bur clover
コウヤカミツレ	*Anthemis tinctoria* L.	
コウライシバ	*Zoysia tenuifolia* Willd.	Korean lawn-grass, Koreangrass, mascarenegrass

コウリンタンポポ	*Hieracium aurantiacum* L.	orange hawkweed
コゴメガヤツリ	*Cyperus iria* L.	yellow-cyperus, rice flatsedge
コゴメギク	*Galinsoga parviflora* Cav.	
コシカギク	*Matricaria matricarioides* Port	
コシナガワハギ	*Melilotus indica* All.	
コスズメノチャヒキ, スムースブロムグラス	*Bromus inermis* Leyss.	smooth brome, smooth bromegrass
コセンダングサ	*Bidens pilosa* L.	
コナスビ	*Lysimachia japonica* Thunb.	loosatrife
コナラ	*Quercus serrata* Thunb.	oak
コニシキソウ	*Euphorbia supina* Rafin.	milk-purslane
コヌカグサ, レッドトップ	*Agrostis alba* L.	redtop
コバギボウシ	*Hosta albo-marginata* (Hook.) Ohwi	narrow leaved plantain lily
コヒメビエ	*Echinochloa colonum* Link	
コヒルガオ	*Calystegia hederacea* Wall.	hede bindweed, larger bindweed, great bindweed, bearbine
コマツナギ	*Indigofera pseudo-tinctoria* Matsum.	indigo
コミカンソウ	*Phyllanthus urinaria* L.	spurge
コムギ	*Triticum aestivum* L.	wheat
コメツブウマゴヤシ	*Medicago lupulina* L.	black medick
コメツブツメクサ	*Trifolium dubium* Sibth.	yellow suckling clover, small hop clover
コメヒシバ	*Digitaria timorensis* (Kunth) Balansa	fingergrass, crabgrass
コモンベッチ, カラスノエンドウ	*Vicia sepium* L.	common vetch
コロニアルベントグラス	*Agrostis tenuis* Sibth.	common bentgrass, colonial bentgrass
コロンブスグラス	*Sorghum almum* Parodi.	Columbus grass
コンフリー	*Symphytum officinale* L.	comfrey
栽培ヒエ	*Echinochloa utilis* Ohwi et Yabuno	Japanese barnyard millet
サイラトロ	*Macroptilium atropurpureum* (DC.) Urb.	siratro
雑色アルファルファ	*Medicago media* Pers.	variegated alfalfa

植物名（和名順）

和名	学名	英名
サツマイモ, カンショ（甘藷）	*Ipomoea batatas* Lam.	sweet potato
サトウキビ	*Saccharum officinarum* L.	sugar cane
サブタレニアンクローバ	*Trifolium subterraneum* L.	subterranean clover, subclover
サルトリイバラ	*Smilax china* L.	greenbrier, catbrier
シオザキソウ	*Tagetes minuta* L.	
シグナルグラス	*Brachiaria decumbens* (Hochst.) Stapf	signalgrass, palisade grass
シコクビエ	*Eleusine coracana* (L.) Gaertn.	African millet
シシウド	*Angelica pubescens* Maxim.	angelica
シナガワハギ, キバナスィートクローバ	*Melilotus officinalis* Pallas	
シナダレスズメガヤ, ウィーピングラブグラス	*Eragrostis curvula* Nees	weeping lovegrass
シバ	*Zoysia japonica* Steud.	Japanese lawngrass
シバスゲ	*Carex nervata* Franch. et Savat.	sedge
シバムギ	*Agropyron repens* (L.) P. Beauv.	quackgrass, dog's-grass, quickgrass
シープフェスク, ウシノケグサ	*Festuca ovina* L.	sheep's fescue
シマスズメノヒエ, ダリスグラス	*Paspalum dilatatum* Poir.	dallisgrass, large watergrass
シマニシキソウ	*Euphorbia hirta* L.	spurge, wolf's milk
ジャイアントスターグラス	*Cynodon aethiopicus* Clayton & Harlan	giant stargrass
ショクヨウガヤツリ, キハマスゲ	*Cyperus esculentus* L.	yellow nutsedge
ジョンソングラス, セイバンモロコシ	*Sorghum halepense* (L.) Pers.	Johnsongrass
シラカンバ	*Betula platyphylla* Sukatchev var. *japonica* (Miq.) Hara	birch
シラゲガヤ, ベルベットグラス	*Holcus lanatus* L.	velvetgrass
シラスゲ	*Carex doniana* Spreng.	sedge
シラヤマギク	*Aster scaber* Thunb.	starwort, frostflower
飼料カブ, レープ	*Brassica rapa* L.	turnip
飼料用ビート	*Beta vulgaris crassa* Alef.	mangold

シルバーリーフデスモディウム	*Desmodium uncinatum* (Jacq.) DC.	silverleaf desmodium, silverleaf Spanish clover
シロクローバ, シロツメクサ	*Trifolium repens* L.	white clover
シロザ	*Chenopodium album* L.	common lambs-quarters, white-goose-foot, green-pigweed
シロツメクサ, シロクローバ	*Trifolium repens* L.	white clover
シロネ	*Lycopus lucidus* Turcz.	
シロバナシナガワハギ, シロバナスィートクローバ	*Melilotus alba* Medic.	white sweetclover
シロバナスィートクローバ, シロバナシナガワハギ	*Melilotus alba* Desr.	white sweetclover
シロバナタンポポ	*Taraxacum albidum* Dahlst.	dandelion, blowballs
シロバナチョウセンアサガオ, ヨウシュチョウセンアサガオ	*Datura stramonium* L.	jimsonweed
シロバナマンテマ	*Silene gallica* L.	small-flowered catchfly, common catchfly
スィートバーナルグラス, ハルガヤ	*Anthoxanthum odoratum* L.	sweet vernalgrass
スイッチグラス	*Panicum virgatum* L.	switchgrass
スイバ	*Rumex acetosa* L.	sorrel
スーダングラス	*Sorghum sudanense* (Piper) Stapf	Sudangrass
スギ	*Cryptomeria japonica* D. Don	
スギナ	*Equisetum arvense* L.	field horsetail
ススキ	*Miscanthus sinensis* Anderss.	silvergrass, Japanese plume-grass
スズタケ	*Sasamorpha borealis* (Hack.) Nakai	
スズメノエンドウ	*Vicia hirsuta* (L.) S.F. Gray	common hairy tare, tyne grass
スズメノカタビラ	*Poa annua* L.	annual bluegrass
スズメノチャヒキ	*Bromus japonicus* Thunb.	Japanese brome
スズメノテッポウ	*Alopecurus aequalis* Sobol.	foxtail

植物名（和名順）

スズメノヒエ	*Paspalum thunbergii* Kunth	knotgrass
スズラン	*Convallaria keiskei* Miq.	lily of the valley, May-lily
スタイロ	*Stylosanthes guianensis* (Aubl.) Sw.	stylo
ストロベリークローバ	*Trifolium fragiferum* L.	strawberry clover
スペアグラス	*Heteropogen contortus* (L.) Beauv.	speargrass
ズミ	*Malus sieboldii* (Regel) Rehder	toringo crab
スムースブロムグラス, コスズメノチャヒキ	*Bromus inermis* Leyss.	smooth bromegrass
スレンダーホィートグラス	*Agropyron trachycautum* Link.	slender wheatgrass
セイタカアワダチソウ	*Solidago altissima* L.	tall goldenrod
セイバンモロコシ, ジョンソングラス	*Sorghum halepense* (L.) Pers.	Johnsongrass
セイヨウアブラナ	*Brassica napus* L.	rape, cole
セイヨウオオバコ	*Plantago major* L.	broad-leaved
セイヨウタンポポ	*Taraxacum officinale* Weber	dandelion
セイヨウトゲアザミ	*Cirsium arvense* Scop.	
セイヨウノコギリソウ	*Achillea millefolium* L.	common yarrow, milfoil
セイヨウノダイコン	*Raphanus raphanistrum* L.	
セイヨウヒルガオ	*Convolvulus arvensis* L.	European bindweed, field bindweed
セイヨウミヤコグサ	*Lotus corniculatus* L.	common birdsfoot trefoil
セタリア	*Setaria sphacelata* (Schumach) Stapf et Hubb.	golden timothy
ゼニバアオイ	*Malva neglecta* Wallr.	
センダングサ	*Bidens biternata* Merr. et Sherff	
センチピードグラス	*Eremochloa ophiuroides* (Munro) Hack.	centipedegrass
セントオーガスチングラス	*Stenotaphrum secundatum* O. Kuntze.	St. Augustine grass
セントロシーマ	*Centrosema pubescents* Benth.	centro butterfly pea
センナリホオズキ	*Physalis angulata* L.	
センブリ	*Swertia japonica* (Schult.) Makino	Columbo
ソバカズラ	*Fallopia convolvulus* (L.) A. Love	
ソルガム, モロコシ	*Sorghum bicolor* Moench	sorghum, sorgo

ダイズ	Glycine max (L.) Merr.	soy bean
タウコギ	Bidens tripartita L.	erect bur marigold
タウンスビルスタイロ	Stylosanthes humilis H. B. K.	townsville stylo, townsville lucerne
タガネソウ	Carex siderosticta Hance	sedge
ダケカンバ	Betula ermanii Cham.	Erman's birch
タケニグサ	Macleaya cordata (Willd.) R. Br.	plume poppy, tree celandine
タチイヌノフグリ	Veronica arvensis L.	corn speedwell, wall speedwell
タチスズメノヒエ, ベージーグラス	Paspalum urvillei Steud.	vaseygrass
タチチチコグサ	Gnaphalium calviceps L.	
タチツボスミレ	Viola grypoceras A. Gray	violet, pansy
タニウツギ	Weigela hortensis (Sieb. et Zucc.) K. Koch	
タヌキマメ	Crotalaria sessiliflora L.	crotalaria
タビラコ	Trigonotis peduncularis (Trevir.) Benth.	
タマガヤツリ	Cyperus difformis L.	umbrella plant, smaller-flower umbrella plant
タラノキ	Aralia elata (Miq.) Seemann	Japanese angelica tree
ダリスグラス, シマスズメノヒエ	Paspalum dilatatum Poir.	dallisgrass, large watergrass
ダンチク	Arundo donax L.	Spanish reed, great reed, bamboo-reed
ダンドボロギク	Erechtites hieracifolia (L.) Rafin.	American burnweed, fire weed
チガヤ	Imperata cylindrica (L.) Beauv. var. koenigii (Retz.) Durand et Schinz.	needlegrass, cogongrass
チカラシバ	Pennisetum alopecuroides (L.) Spreng.	
チゴザサ	Isachne globosa (Thunb.) O. Kuntze	
チゴユリ	Disporum smilacinum A. Gray	fairybell
チシマザサ	Sasa kurilensis (Rupr.) Makino et Shibata	sasa, dwarf bamboo
チダケサシ	Astilbe microphylla Knoll	false goatsbeard
チチコグサ	Gnaphalium japonicum Thunb.	cudweed, everlasting

植物名（和名順）

チヂミザサ	*Oplismenus undulatifolius* (Ard.) Roemer et Schultes var. *japonicus* (Steud.) Koidz.	
チドメグサ	*Hydrocotyle sibthorpioides* Lam.	lawn pennywort, shining pennywort
チマキザサ	*Sasa palmata* (Marliac) Nakai	sasa, dwarf bamboo
チモシー, オオアワガエリ	*Phleum pratense* L.	timothy
チョウセンアサガオ	*Datura metel* L.	
ツボミオオバコ	*Plantago virginica* L.	white dwarf plantain, bearded plantain
ツメクサ	*Sagina japonica* (Sw.) Ohwi	pearlwort
ツユクサ	*Commelina communis* L.	Asiatic dayflower
ツリガネニンジン	*Adenophora triphylla* (Thunb.) A. DC. var. *japonica* (Regel) Hara	
ツルマメ	*Glycine soja* Sieb. et Zucc.	
ディジットグラス, パンゴラグラス, フィンガーグラス	*Digitaria eriantha* Steud.	digitgrass, pangolagrass, fingergrass
テオシント	*Euchlaena mexicana* Schrad.	teosinte
テフ	*Eragrostis abyssinica* (Jacq.) Link	teff, teff grass
デリーグラス, マーベルグラス	*Dichanthium annulatum* (Forsk.) Stapf	marvelgrass, sheda grass, Delhi grass
テリハノイバラ	*Rosa wichuraiana* Crep	memorial rose
テンニンソウ	*Leucosceptrum japonicum* (Miq.) Kitam. et Murata	
トウジンビエ, パールミレット	*Pennisetum typhoides* (Burmf.) Stapf	pearl millet
トウダイグサ	*Euphorbia helioscopia* L.	devil's-milk, wortweed, sun spurge
トウモロコシ	*Zea mays* L.	corn, maize
トールオートグラス, オオカニツリ	*Arrhenatherum elatius* (L.) Mertens et Koch	tall oatgrass, false oatgrass
トールフェスク, オニウシノケグサ	*Festuca arundinacea* Schreb.	tall fescue
トキワススキ	*Miscanthus floridulus* (Labill.) Warb.	eulalia
トキワハゼ	*Mazus pumilus* (Burm. fil.) V. Steenis	

ドクニンジン	*Conium maculatum* L.	
トダシバ	*Arundinella hirta* (Thunb.) C. Tanaka	
トボシガラ	*Festuca parvigluma* Steud.	fescue-grass
トリアシショウマ	*Astilbe thunbergii* (Sieb.et Zucc.) Miq. var. *congesta* H. Boiss.	false goatsbeard
トリカブト	*Aconitum chinense* Sieb. et Paxt.	
ナガバギシギシ	*Rumex crispus* L.	curly dock
ナガハグサ, ケンタッキーブルーグラス	*Poa pratensis* L.	Kentucky bluegrass
ナガボノウルシ	*Sphenoclea zeylanica* Gaertn.	
ナギナタガヤ	*Festuca myuros* (L.) C. Gmelin	rat's-tail
ナズナ	*Capsella bursa-pastoris* (L.) Medic.	Shepherds purse
ナルコスゲ	*Carex curvicollis* Franch. et Savat.	sedge
ナルコビエ	*Eriochloa villosa* (Thunb.) Kunth	
ナワシロイチゴ	*Rubus parvifolius* L.	bramble
ナンブアザミ	*Cirsium nipponicum* (Maxim.) Makino	common thistle, plumed thistle
ニガイチゴ	*Rubus microphyllus* L. fil.	bramble
ニガナ	*Ixeris dentata* (Thunb.) Nakai	
ニシキソウ	*Euphorbia pseudochamaesyce* Fisch.	spurge, wolf's milk
ニワホコリ	*Eragrostis multicaulis* Steud.	lovegrass
ヌカキビ	*Panicum bisulcatum* Thunb.	panicum
ヌカボ	*Agrostis clavata* Trin. var. *nukabo* Ohwi	bentgrass
ヌスビトハギ	*Desmodium oxyphyllum* DC.	
ネコハギ	*Lespedeza pilosa* (Thunb.) Sieb. et Zucc.	bush-clover
ネザサ	*Pleioblastus chino* (Franch. et Savat.) Makino	
ネジバナ	*Spiranthes sinensis* (Pers.) Ames	lady's-tresses, pearl-twist
ネズミガヤ	*Muhlenbergia japonica* Steud.	muhly
ネズミノオ	*Sporobolus fertilis* (Steud.) W. Clayton	smutgrass
ネズミムギ, イタリアンライグラス	*Lolium multiflorum* Lam.	Italian ryegrass
熱帯クズ	*Pueraria phaseoloides* (Roxb.) Benth.	tropical kudzu

植物名（和名順）

和名	学名	英名
ネピアグラス	*Pennisetum purpureum* Schum.	napiergrass, elephantgrass
ノアザミ	*Cirsium japonicum* DC.	common thistle, plumed thistle
ノアズキ	*Dunbaria villosa* (Thunb.) Makino	
ノイバラ	*Rosa multiflora* Thunb.	polyantha rose
ノガリヤス	*Calamagrostis arundinacea* (L.) Roth var. *brachytricha* (Steud.) Hack.	small-reed
ノカンゾウ	*Hemerocallis longituba* Miq.	day-lily
ノゲシ	*Sonchus oleraceus* L.	sow-thistle
ノコンギク	*Aster ageratoides* Turcz. var. *ovatus* (Franch. et Savat.) Nakai	starwort, frost-flower
ノジアオイ	*Melochia corchorifolia* L.	
ノダケ	*Angelica decursiva* (Miq.) Franch. et Savat.	angelica
ノチドメ	*Hydrocotyle maritima* Honda	water-pennywort, navelwort
ノハナショウブ	*Iris ensata* Thunb. var. *spontanea* (Makino) Nakai	iris
ノハラアザミ	*Cirsium tanakae* (Franch. et Savat.) Matsum.	common thistle, plumed thistle
ノハラツメクサ	*Spergula arvensis* L.	
ノビル	*Allium grayi* Regel	garlic
ノミノツヅリ	*Arenaria serpyllifolia* L.	thyme-leaved sandwort
ノミノフスマ	*Stellaria alsine* Grimm var. *undulata* (Thunb.) Ohwi	slender slarwort bog stitchwort
ノラニンジン	*Daucus carota* L.	wild carrot
ノリウツギ	*Hydrangea paniculata* Siebold	hydrangea
バークローバ, ウマゴヤシ	*Medicago polymorpha* L.	bur clover
バーズフットトレフォイル, ミヤコグサ	*Lotus corniculatus* L. var. *japonicus* Regel	birdsfoot trefoil
ハーディンググラス	*Phalaris tuberosa* L. var. *stenoptera* (Hack.) Hitchc.	hardinggrass
バーミューダグラス, ギョウギシバ	*Cynodon dactylon* (L.) Pers.	bermudagrass

パールミレット, トウジンビエ	*Pennisetum typhoides* (Burmf.) Stapf et Hubb.	pearl millet
ハイキビ	*Panicum repens* L.	torpedograss
ハイコヌカグサ, クリーピングベントグラス	*Agrostis stolonifera* L.	creeping bentgrass
ハキダメギク	*Galinsoga ciliata* (Raf.) Blake.	
ハコベ	*Stellaria media* (L.) Villars	chickweed
ハシリドコロ	*Scopolia japonica* Maxim.	
ハチジョウススキ	*Miscanthus condensatus* Hack.	eulalia
バッコヤナギ	*Salix bakko* Kimura	willow
バッフェルグラス	*Cenchrus ciliaris* L.	buffelgrass, African foxtail
ハトムギ	*Coix lacryma-jobi* L. var. *mayuen* (Roman.) Stapf	Job's tear
ハネガヤ	*Achantherum extremiorientale* (Hara) Keng	feather-grass
ハハコグサ	*Gnaphalium affine* D. Don	cudweed, everlasting
バヒアグラス	*Paspalum notatum* Flugge	bahiagrass
ハブソウ	*Cassia occidentalis* L.	
ハマチャヒキ	*Bromus mollis* L.	soft chess
ハマナス	*Rosa rugosa* Thunb.	rose
パラグラス	*Brachiaria mutica* (Forsk.) Stapf	paragrass
ハリエンジュ	*Robinia pseudo-acacia* L.	black locust
ハリビユ	*Amaranthus spinosus* L.	thorny amaranth
ハルガヤ, スィートバーナルグラス	*Anthoxanthum odoratum* L.	sweet vernalgrass
ハルザキヤマガラシ	*Barbarea vulgaris* R.Br.	
ハルジオン	*Erigeron philadelphicus* L.	common fleabane, Philadelphia fleabane, fleabane
ハルタデ	*Polygonum persicaria* L.	persicary, red-legs, peachwort, lady's-thumb
ハルリンドウ	*Gentiana thunbergii* (G. Don) Griseb.	gentian
パンゴラグラス, ディジットグラス, フィンガーグラス	*Digitaria eriantha* Steud.	pangolagrass, digitgrass, fingergrass
ハンゴンソウ	*Senecio cannabifolius* Less.	groundsel, ragwort, squaw-weed
パンパスグラス	*Cortaderia argentea* Stapf	pampasgrass

植物名（和名順）

和名	学名	英名
ヒカゲスゲ	*Carex floribunda* (Korsh.) Meinsh.	sedge
ヒゲノガリヤス	*Calamagrostis longiseta* Hack.	small-reed
ビッグブルーステム	*Andropogon gerardii* Vitmen.	big bluestem
ヒマワリ	*Helianthus annuus* L.	sunflower
ヒメオトギリ	*Hypericum japonicum* Thunb.	matted St. John's wort
ヒメオドリコソウ	*Lamium purpureum* L.	purple-dead-nettle, red deadnettle
ヒメクグ	*Cyperus brevifolius* (Rottb.) Hassk. var. *leiolepis* (Fr. et Sav.) T. Koyama	
ヒメジョオン	*Erigeron annuus* (L.) Pers.	annual fleabane, white-top, daisy fleabane, sweet scabious
ヒメスイバ	*Rumex acetosella* L.	red sorrel
ヒメミソハギ	*Ammannia multiflora* Roxb.	red stem
ヒメムカシヨモギ	*Erigeron canadensis* L.	horseweed, hog-weed, butterweed
ヒメヤブラン	*Liriope minor* (Maxim.) Makino	lily-turf
ヒヨドリバナ	*Eupatorium chinense* L. var. *simplicifolium* (Makino) Kitam.	thoroughwort
ヒルガオ	*Calystegia japonica* Choisy	
ヒレアザミ	*Carduus crispus* L.	curled thistle, welted thistle
ヒレタゴボウ	*Ludwigia decurrens* Walt.	
ビロードモウズイカ	*Verbascum thapsus* L.	
ファジービーン	*Macroptilium lathyroides* (L.) Urb.	phasey bean
フィンガーグラス, パンゴラグラス, ディジットグラス	*Digitaria eriantha* Steud.	fingergrass, pangolagrass, digitgrass
フェストロリウム	× *Festulolium* Braunii	
フキ	*Petasites japonicus* (Sieb. et Zucc.) Maxim.	lagwort
フジ	*Wistaria floribunda* (Willd.) DC.	Japanese wistaria
ブタクサ	*Ambrosia artemisiiflolia* L. var. *elatior* Desc.	common ragweed, Roman ragweed, hogweed
ブタクサモドキ	*Ambrosia psilostachya* DC.	

ブタナ	*Hypochoeris radicata* L.	
フデリンドウ	*Gentiana zollingeri* Fawcett	gentian
ブナ	*Fagus crenata* Blume	Siebold's beech
フラサバソウ	*Veronica hederaefolia* L.	
フランスギク	*Chrysanthemum leucanthemum* L.	
ブルーグラマ	*Bouteloua gracilis* Lag.	blue grama
ブルーパニック	*Panicum antidotale* Retz.	blue panic
ヘアリーベッチ	*Vicia villosa* Roth	hairy vetch
ベージーグラス, タチスズメノヒエ	*Paspalum urvillei* Steud.	vaseygrass
ヘクソカズラ	*Paederia scandens* (Lour.) Merrill var. *mairei* (Léveille) Hara	
ベニバナボロギク	*Crassocephalum crepidioides* (Benth.) S. Moore	
ヘビイチゴ	*Duchesnea chrysantha* (Zoll. et Mor.) Miq.	India mockstrawberry, wild strawberry
ヘラオオバコ	*Plantago lanceolata* L.	buckthorn plantain, ribbed plantain, English plantain
ヘラバヒメジョオン	*Erigeron strigosus* Muhl.	
ベルベットグラス, シラゲガヤ	*Holcus lanatus* L.	velvetgrass
ペレニアルライグラス, ホソムギ	*Lolium perenne* L.	perennial ryegrass
ホウキギク	*Aster subulatus* Michx.	annual saltmarsh aster
ホシアサガオ	*Ipomoea triloba* L.	
ホソアオゲイトウ	*Amaranthus patulus* Bertol.	
ホソバヒメミソハギ	*Ammannia coccinea* Rottb.	red stem, long-leaved ammannia
ホソムギ, ペレニアルライグラス	*Lolium perenne* L.	perennial ryegrass
ホトケノザ	*Lamium amplexicaule* L.	henbit, perfoliate-archangel henbit, deadnettle
ホナガアオゲイトウ	*Amaranthus hybridus* L.	
マーベルグラス, デリーグラス	*Dichanthium annulatum* (Forsk.) Stapf	marvelgrass, sheda grass, Delhi grass
マウンテンブロムグラス	*Bromus marginatus* Nees	mountain bromegrass
マカリカリグラス	*Panicum coloratum* L. var. *makarikariense* Goossens	makarikarigrass

植物名（和名順）

マツバウンラン	*Linaria canadensis* Dum.	
マツバセリ	*Apium leptophyllum* (Pers.) F. Muell.	
マツムシソウ	*Scabiosa japonica* Miq.	scabious
マツヨイグサ	*Oenothera striata* Ledeb.	
マムシグサ	*Arisaema japonicum* Blume	Indian-turnip, dragon-arum
マメアサガオ	*Ipomoea lacunosa* L.	small-flowered white morning-glory
マメグンバイナズナ	*Lepidium virginicum* L.	poor-man's pepper
マルバアサガオ	*Ipomoea purpurea* (L.) Roth	
マルバアメリカアサガオ	*Ipomoea hederacea* (L.) Jacq. var *integriuscula* A. Gray	
マルバハギ	*Lespedeza cyrtobotrya* Miq.	bush-clover
マルバルコウ	*Ipomoea coccinea* L.	scarlet morningglory
マンテマ	*Silene gallica* L. var. *quinquevulnera* (L.) Rohrb.	campion
ミズナラ	*Quercus mongolica* Fischer var. *grosseserrata* (Blume) Rehd. et Wils.	oak
ミチヤナギ	*Polygonum aviculare* L.	blood-wort, bird's-knotgrass centynody, yard knotweed
ミツバツチグリ	*Potentilla freyniana* Bornm.	cinquefoil, potentil, five-finger
ミナトアカザ	*Chenopodium murale* L.	
ミノボロスゲ	*Carex albata* Boott	sedge
ミミナグサ	*Cerastium holosteoides* Fries var. *angustifolium* (Franch) Mizushima	mouse-ear, chickweed
ミヤコグサ，バーズフットトレフォイル	*Lotus corniculatus* L. var. *japonicus* Regel	birdsfoot trefoil
ミヤコザサ	*Sasa nipponica* (Makino) Makino et Shibata	sasa, dwarf bamboo
ムラサキウマゴヤシ，アルファルファ，ルーサン	*Medicago sativa* L.	alfalfa, lucerne
ムラサキサギゴケ	*Mazus miquelii* Makino	
ムラサキヒゲシバ	*Chloris barbata* Swartz	purple top chloris
メーヤーデスモディウム	*Desmodium sandwicense* E. Meyer	Spanish clover, sandwitch

メガルカヤ	*Themeda japonica* (Willd.) C. Tanaka	
メドウフェスク	*Festuca pratensis* Huds.	meadow fescue
メドウフォクステイル	*Alopecurus pratensis* L.	meadow foxtail
メドウブロムグラス	*Bromus erectus* Huds.	meadow brome, meadow bromegrass
メドハギ	*Lespedeza cuneata* (Du Mont. d. Cours.) G.Don	sericea lespedeza
メナモミ	*Siegesbeckia pubescens* (Makino) Makino	
メヒシバ	*Digitaria adscendens* (H.B.K.) Henr.	southern crabgrass, Henry crabgrass, hairy crabgrass
メマツヨイグサ	*Oenothera biennis* L.	common evening primrose
メリケンカルカヤ	*Andropogon virginicus* L.	broomsedge
メリケンニクキビ	*Brachiaria platyphylla* Nash	
モウズイカ	*Verbascum blattaria* L.	
モミジイチゴ	*Rubus palmatus* Thunb. var. *coptophyllus* (A. Gray) Koidz.	bramble
モラセスグラス	*Melinis minutiflora* Beauv.	molasses grass
モロコシ, ソルガム	*Sorghum bicolor* Moench	sorghum, sorgo
モンツキウマゴヤシ	*Medicago arabica* (L.) Huds.	
ヤエムグラ	*Galium spurium* L. var. *echinospermon* (Wallr.) Hayek	bedstraw
ヤクシソウ	*Youngia denticulata* (Houtt.) Kitam.	
ヤハズソウ	*Lespedeza striata* (Thunb.) Schindler	common lespedeza, Japan clover
ヤブカンゾウ	*Hemerocallis fulva* L. var. *kwanso* Regel	day-lily
ヤブマメ	*Amphicarpaea edgeworthii* Benth. var. *japonica* Oliver	hog-peanut
ヤマアジサイ	*Hydrangea macrophylla* (Thunb.) Ser. var. *acuminata* (Sieb. et Zucc.) Makino	hydrangea
ヤマアワ	*Calamagrostis epigeios* (L.) Roth	small-reed, wood-reed, bush-grass
ヤマジノホトトギス	*Tricyrtis affinis* Makino	Japanese toad-lily

植物名（和名順）

ヤマシロギク	*Aster sugimotoi* Kitamura var. *semiamplexicaulis* (Makino) Ohwi	starwort, frost-flower
ヤマツツジ	*Rhododendron kaempferi* Planch.	azalea
ヤマドリゼンマイ	*Osmundastrum cinnamomeum* (L.) Pr. var. *fokiense* (Copel) Tagawa	
ヤマナラシ	*Populus sieboldii* Miq.	Japanese aspen
ヤマヌカボ	*Agrostis clavata* Trin.	bentgrass
ヤマハンノキ	*Alnus hirsuta* Turcz. var. *sibirica* (Fischer) C.K. Schn.	alder
ヤマブキショウマ	*Aruncus dioicus* (Walt.) Fernald var. *tenuifolius* (Nakai) Hara	goat's-beard
ヤマモミジ	*Acer palmatum* Thunb. var. *matumurae* (Koidz.) Makino	Japanese maple
ヤマユリ	*Lilium auratum* Lindl.	goldband lily, golden lily
ヤリテンツキ	*Fimbristylis ovata* (Burm fil.) Kern	
ユウガギク	*Kalimeris pinnatifida* (Maxim.) Kitam.	
ヨウシュチョウセンアサガオ, シロバナチョウセンアサガオ	*Datura stramonium* L.	jimsonweed
ヨシ	*Phragmites communis* Trin.	common reed
ヨメナ	*Kalimeris yomena* Kitam.	
ヨモギ	*Artemisia princeps* Pampan.	mugwort
ライコムギ	× *Triticosecale* Wittmack	triticale
ライムギ	*Secale cereale* L.	rye
リードカナリーグラス, クサヨシ	*Phalaris arundinacea* L.	reed canarygrass
リトルブルーステム	*Andropogon scoparius* Michx.	little bluestem
リンドウ	*Gentiana scabra* Bunge var. *buergeri* (Miq.) Maxim.	gentian
リンポグラス	*Hemarthria altissima* (Poir.) Stapf et Hubbard	limpograss
ルーサン, ムラサキウマゴヤシ, アルファルファ	*Medicago sativa* L.	alfalfa, lucerne
ルーピン（青花）	*Lupinus angustifolius* L.	blue lupin
ルタバガ	*Brassica napus* L. subsp. *rapifera* Metzg. Sink.	rutabaga

レープ, 飼料カブ	*Brassica rapa* L.	turnip
レスクグラス, イヌムギ	*Bromus catharticus* Vahl	rescuegrass
レッドトップ, コヌカグサ	*Agrostis alba* L.	redtop
レッドフェスク, クリーピングフェスク, オオウシノケグサ	*Festuca rubra* L.	red fescue, creeping fescue
レンゲ, ゲンゲ	*Astragalus sinicus* L.	Chinese milk vetch
レンゲツツジ	*Rhododendron japonicum* (A. Gray) Suringer	azalea
ローズグラス	*Chloris gayana* Kunth	rhodesgrass
ローマカミツレ	*Anthemis nobilis* L.	
ワラビ	*Pteridium aquilinum* (L.) Kuhn var. *latiusculum* (Desv.) Und.	eastern bracken, brackenfern
ワルナスビ	*Solanum carolinense* L.	horse nettle
ワレモコウ	*Sanguisorba officinalis* L.	burnet-blood wort

植物名 （学名順）

学 名	和 名	英 名
Abutilon theophrasti Medic.	イチビ	velvetleaf, piemaker
Acer palmatum Thunb. var. *matumurae* (Koidz.) Makino	ヤマモミジ	Japanese maple
Achantherum extremiorientale (Hara) Keng	ハネガヤ	feather-grass
Achillea millefolium L.	セイヨウノコギリソウ	common yarrow, milfoil
Aconitum chinense Sieb. et Paxt.	トリカブト	
Adenophora triphylla (Thunb.) A. DC. var. *japonica* (Regel) Hara	ツリガネニンジン	
Ageratum houstonianum Mill.	オオカッコウアザミ	
Agrimonia japonica (Miq.) Koidz.	キンミズヒキ	agrimony, cocklebur
Agropyron cristatum Gaertn.	クレステッドホィートグラス	crested wheatgrass
Agropyron intermedium P. Beauv.	インターミーディエトホィートグラス	intermediate wheatgrass
Agropyron racemiferum (Steud.) Koidz.	アオカモジグサ	
Agropyron repens (L.) P. Beauv.	シバムギ	quackgrass, dog's-grass, quick-grass
Agropyron trachycautum Link.	スレンダーホィートグラス	slender wheatgrass
Agropyron tsukushiense (Honda) Ohwi var. *transiens* (Hack.) Ohwi	カモジグサ	drooping wheatgrass
Agrostis alba L.	レッドトップ, コヌカグサ	redtop
Agrostis capillaris L.	イトコヌカグサ	
Agrostis clavata Trin.	ヤマヌカボ	bentgrass
Agrostis clavata Trin. var. *nukabo* Ohwi	ヌカボ	bentgrass
Agrostis stolonifera L.	ハイコヌカグサ, クリーピングベントグラス	creeping bentgrass
Agrostis tenuis Sibth.	コロニアルベントグラス	common bentgrass, colonial bentgrass

Ajuga decumbens Thunb.	キランソウ	bugle
Allium grayi Regel	ノビル	garlic
Alnus hirsuta Turcz. var. *sibirica* (Fischer) C.K. Schn.	ヤマハンノキ	alder
Alopecurus aequalis Sobol.	スズメノテッポウ	foxtail
Alopecurus pratensis L.	メドウフォクステイル	meadow foxtail
Alysicarpus vaginalis (L.) DC.	アリスクローバ	alyce clover
Amaranthus hybridus L.	ホナガアオゲイトウ	
Amaranthus lividus L.	イヌビユ	livid amaranth
Amaranthus palmeri S. Wats.	オオホナガアオゲイトウ	
Amaranthus patulus Bertol.	ホソアオゲイトウ	
Amaranthus retroflexus L.	アオゲイトウ	green amaranth
Amaranthus spinosus L.	ハリビユ	thorny amaranth
Amaranthus viridis L.	アオビユ	slender amaranth
Ambrosia artemisiifolia L. var. *elatior* Desc.	ブタクサ	common ragweed, Roman ragweed, hogweed
Ambrosia psilostachya DC.	ブタクサモドキ	
Ambrosia trifida L.	オオブタクサ	giant ragweed
Ammannia coccinea Rottb.	ホソバヒメミソハギ	red stem, long-leaved ammannia
Ammannia multiflora Roxb.	ヒメミソハギ	red stem
Amphicarpaea edgeworthii Benth. var. *japonica* Oliver	ヤブマメ	hog-peanut
Andropogon brevifolius Swartz	ウシクサ	beardgrass
Andropogon gerardii Vitmen.	ビッグブルーステム	big bluestem
Andropogon scoparius Michx.	リトルブルーステム	little bluestem
Andropogon virginicus L.	メリケンカルカヤ	broomsedge
Angelica decursiva (Miq.) Franch. et Savat.	ノダケ	angelica
Angelica pubescens Maxim.	シシウド	angelica
Angelica ursina (Rupr.) Maxim.	エゾニュウ	angelica
Anthemis arvensis L.	キゾメカミツレ	
Anthemis cotula L.	カミツレモドキ	
Anthemis nobilis L.	ローマカミツレ	
Anthemis tinctoria L.	コウヤカミツレ	
Anthoxanthum odoratum L.	ハルガヤ, スィートバーナルグラス	sweet vernalgrass
Apium leptophyllum (Pers.) F. Muell.	マツバセリ	
Aralia cordata Thunb.	ウド	angelica tree

Aralia elata (Miq.) Seemann	タラノキ	Japanese angelica tree
Arenaria serpyllifolia L.	ノミノツヅリ	thyme-leaved sandwort
Arisaema japonicum Blume	マムシグサ	Indian-turnip, dragon-arum
Arrhenatherum elatius (L.) Mertens et Koch	トールオートグラス, オオカニツリ	tall oatgrass, false oatgrass
Artemisia japonica Thunb.	オトコヨモギ	wormwood
Artemisia princeps Pampan.	ヨモギ	mugwort
Aruncus dioicus (Walt.) Fernald var. *tenuifolius* (Nakai) Hara	ヤマブキショウマ	goat's-beard
Arundinella hirta (Thunb.) C. Tanaka	トダシバ	
Arundo donax L.	ダンチク	Spanish reed, great reed, bamboo-reed
Aster ageratoides Turcz. var. *semiamplexicaulis* (Makino) Ohwi	ヤマシロギク	starwort, frost-flower
Aster scaber Thunb.	シラヤマギク	starwort, frostflower
Aster subulatus Michx.	ホウキギク	annual saltmarsh aster
Aster sugimotoi Kitamura var. *ovatus* (Franch. et Savat.) Nakai	ノコンギク	starwort, frost-flower
Astilbe microphylla Knoll	チダケサシ	false goatsbeard
Astilbe thunbergii (Sieb. et Zucc.) Miq. var. *congesta* H. Boiss.	トリアシショウマ	false goatsbeard
Astragalus sinicus L.	ゲンゲ, レンゲ	Chinese milk vetch
Avena fatua L.	カラスムギ	wild oat
Avena sativa L.	エンバク	oat
Axonopus affinis Chase	カーペットグラス	carpetgrass
Bacopa rotundifolia Wettst.	ウキアゼナ	disc waterhyssop
Barbarea vulgaris R.Br.	ハルザキヤマガラシ	
Beta vulgaris crassa Alef.	飼料用ビート	mangold
Betula ermanii Cham.	ダケカンバ	Erman's birch
Betula platyphylla Sukatchev var. *japonica* (Miq.) Hara	シラカンバ	birch
Bidens biternata Merr. et Sherff	センダングサ	

植物名（学名順）

Bidens frondosa L.	アメリカセンダングサ	devils beggar-ticks, beggar-ticks, stick-tight
Bidens pilosa L.	コセンダングサ	
Bidens tripartita L.	タウコギ	erect bur marigold
Boehmeria tricuspis (Hance) Makino	アカソ	false nettle
Bouteloua gracilis Lag.	ブルーグラマ	blue grama
Brachiaria decumbens (Hochst.) Stapf	シグナルグラス	*signalgrass,* palisade grass
Brachiaria mutica (Forsk.) Stapf	パラグラス	paragrass
Brachiaria platyphylla Nash	メリケンニクキビ	
Brassica napus L.	セイヨウアブラナ	rape, cole
Brassica napus L. subsp. *rapifera* Metzg. Sink.	ルタバガ	rutabaga
Brassica rapa L.	飼料カブ，レープ	turnip
Bromus catharticus Vahl	イヌムギ，レスクグラス	rescuegrass
Bromus erectus Huds.	メドウブロムグラス	meadow brome, meadow bromegrass
Bromus inermis Leyss.	スムースブロムグラス，コスズメノチャヒキ	smooth bromegrass
Bromus japonicus Thunb.	スズメノチャヒキ	Japanese brome
Bromus marginatus Nees	マウンテンブロムグラス	mountain bromegrass
Bromus mollis L.	ハマチャヒキ	soft chess
Bromus sterilis L.	アレチノチャヒキ	
Bromus tectorum L.	ウマノチャヒキ	downy brome
Calamagrostis arundinacea (L.) Roth var. *brachytricha* (Steud.) Hack.	ノガリヤス	small-reed
Calamagrostis epigeios (L.) Roth	ヤマアワ	small-reed, wood-reed, bush-grass
Calamagrostis langsdorffii (Link) Trin	イワノガリヤス	small-reed
Calamagrostis longiseta Hack.	ヒゲノガリヤス	small-reed
Calystegia hederacea Wall.	コヒルガオ	hede bindweed, larger bindweed, great bindweed, bearbine
Calystegia japonica Choisy	ヒルガオ	

Capsella bursa-pastoris (L.) Medic.	ナズナ	Shepherds purse
Carduus crispus L.	ヒレアザミ	curled thistle, welted thistle
Carex albata Boott	ミノボロスゲ	sedge
Carex curvicollis Franch. et Savat.	ナルコスゲ	sedge
Carex doniana Spreng.	シラスゲ	sedge
Carex floribunda (Korsh.) Meinsh.	ヒカゲスゲ	sedge
Carex kobomugi Ohwi	コウボウムギ	sedge
Carex nervata Franch. et Savat.	シバスゲ	sedge
Carex siderosticta Hance	タガネソウ	sedge
Cassia nomame (Sieb.) Honda	カワラケツメイ	senna
Cassia obtusifolia L.	エビスグサ	sicklepod
Cassia occidentalis L.	ハブソウ	
Castanea crenata Sieb. et Zucc.	クリ	chestnut
Cenchrus brownii Roem. et Schult.	クリノイガ	
Cenchrus ciliaris L.	バッフェルグラス	buffelgrass, African foxtail
Centrosema pubescents Benth.	セントロシーマ	centro butterfly pea
Cerastium glomeratum Thuill.	オランダミミナグサ	sticky mouse-ear, chickweed
Cerastium holosteoides Fries var. *angustifolium* (Franch) Mizushima	ミミナグサ	mouse-ear, chickweed
Chenopodium album L.	シロザ	common lambsquarters, white-goose-foot, green foot
Chenopodium album L. var. *centrorubrum* Makino	アカザ	goose-foot pigweed
Chenopodium carinatum R. Br.	ゴウシュウアリタソウ	
Chenopodium ficifolium Smith	コアカザ	fig-leaved goose foot
Chenopodium murale L.	ミナトアカザ	
Chloris barbata Swartz	ムラサキヒゲシバ	purple top chloris
Chloris gayana Kunth	ローズグラス	rhodesgrass
Chloris virgata Swartz	オヒゲシバ	feather top rhodesgrass, feather fingergrass

Chrysanthemum leucanthemum L.	フランスギク	
Chrysopogon aciculatus (Rstz.) Trin.	オキナワミチシバ	
Cirsium arvense Scop.	セイヨウトゲアザミ	
Cirsium japonicum DC.	ノアザミ	common thistle, plumed thistle
Cirsium nipponicum (Maxim.) Makino	ナンブアザミ	common thistle, plumed thistle
Cirsium tanakae (Franch. et Savat.) Matsum.	ノハラアザミ	common thistle, plumed thistle
Cirsium vulgare (Savi) Tenore	アメリカオニアザミ	bull thistle
Cocculus orbiculatus (L.) Forman	アオツヅラフジ	coralbeads
Coix lacryma-jobi L. var. *mayuen* (Roman.) Stapf	ハトムギ	Job's tear
Commelina communis L.	ツユクサ	Asiatic dayflower
Conium maculatum L.	ドクニンジン	
Convallaria keiskei Miq.	スズラン	lily of the valley, May-lily
Convolvulus arvensis L.	セイヨウヒルガオ	European bindweed, field bindweed
Coronopus didymus Smith	カラクサガラシ, カラクサナズナ	
Cortaderia argentea Stapf	パンパスグラス	pampasgrass
Crassocephalum crepidioides (Benth.) S. Moore	ベニバナボロギク	
Crotalaria juncea L.	クロタラリア	sun hemp
Crotalaria sessiliflora L.	タヌキマメ	crotalaria
Cryptomeria japonica D. Don	スギ	
Cuscuta pentagona Engelm.	アメリカネナシカズラ	
Cymbopogon tortilis (Presl) A. Camus var. *goeringii* (Steud.) Hand.-Mazz.	オガルカヤ	
Cynodon aethiopicus Clayton & Harlan	ジャイアントスターグラス	giant stargrass
Cynodon dactylon (L.) Pers.	バーミューダグラス, ギョウギシバ	bermudagrass
Cyperus brevifolius (Rottb.) Hassk. var. *leiolepis* (Fr. et Sav.) T. Koyama	ヒメクグ	

Cyperus difformis L.	タマガヤツリ	umbrella plant, smaller-flower umbrella plant
Cyperus esculentus L.	ショクヨウガヤツリ, キハマスゲ	yellow nutsedge
Cyperus globosus All.	アゼガヤツリ	
Cyperus iria L.	コゴメガヤツリ	yellow-cyperus, rice flatsedge
Cyperus microiria Steud.	カヤツリグサ	chufa, flatsedge, umbrella sedge
Dactylis glomerata L.	オーチャードグラス, カモガヤ	orchardgrass, cocksfoot
Datura metel L.	チョウセンアサガオ	
Datura stramonium L.	シロバナチョウセンアサガオ, ヨウシュチョウセンアサガオ	jimsonweed
Daucus carota L.	ノラニンジン	wild carrot
Desmodium intortum (Mill.) Urb.	グリーンリーフデスモディウム	greenleaf desmodium, kuru vine
Desmodium oxyphyllum DC.	ヌスビトハギ	
Desmodium sandwicense E. Meyer	メーヤーデスモディウム	Spanish clover, sandwitch
Desmodium uncinatum (Jacq.) DC.	シルバーリーフデスモディウム	silverleaf desmodium, silverleaf Spanish clover
Dianthus superbus L. var. *longicalycinus* (Maxim.) Williams	カワラナデシコ	superb-pink
Dichanthium annulatum (Forsk.) Stapf	デリーグラス, マーベルグラス	marvelgrass, sheda grass, Delhi grass
Digitaria adscendens (H.B.K.) Henr.	メヒシバ	southern crabgrass, Henry crabgrass, hairy crabgrass
Digitaria eriantha Steud.	ディジットグラス, フィンガーグラス, パンゴラグラス	digitgrass, fingergrass, pangolagrass
Digitaria timorensis (Kunth) Balansa	コメヒシバ	fingergrass, crabgrass
Digitaria violascens Link	アキメヒシバ	small crabgrass, violet crabgrass
Disporum smilacinum A. Gray	チゴユリ	fairybell

Duchesnea chrysantha (Zoll. et Mor.) Miq.	ヘビイチゴ	India mockstrawberry, wild strawberry
Dunbaria villosa (Thunb.) Makino	ノアズキ	
Eccoilopus cotulifer (Thunb.) A. Camus	アブラススキ	
Echinochloa colonum Link	コヒメビエ	
Echinochloa crus-galli (L.) Beauv.	イヌビエ	barnyardgrass
Echinochloa utilis Ohwi et Yabuno	栽培ヒエ	Japanese barnyard millet
Elaeagnus umbellata Thunb.	アキグミ	oleaster
Eleusine coracana (L.) Gaertn.	シコクビエ	African millet
Eleusine indica (L.) Gaertn.	オヒシバ	goosegrass
Equisetum arvense L.	スギナ	field horsetail
Eragrostis abyssinica (Jacq.) Link	テフ	teff, teff grass
Eragrostis curvula Nees	ウィーピングラブグラス, シナダレスズメガ	weeping lovegrass
Eragrostis ferruginea (Thunb.) Beauv.	カゼクサ	lovegrass
Eragrostis multicaulis Steud.	ニワホコリ	lovegrass
Erechtites hieracifolia (L.) Rafin.	ダンドボロギク	American burnweed, fire weed
Eremochloa ophiuroides (Munro) Hack.	センチピードグラス	centipedegrass
Erigeron annuus (L.) Pers.	ヒメジョオン	annual fleabane, white-top, daisy fleabane, sweet scabious
Erigeron bonariensis L.	アレチノギク	
Erigeron canadensis L.	ヒメムカシヨモギ	horseweed, hog-weed, butterweed
Erigeron philadelphicus L.	ハルジオン	common fleabane, Philadelphia fleabane, fleabane
Erigeron strigosus Muhl.	ヘラバヒメジョオン	
Erigeron sumatrensis Retz.	オオアレチノギク	
Erigeron thunbergii A. Gray	アズマギク	

Eriochloa villosa (Thunb.) Kunth	ナルコビエ	
Euchlaena mexicana Schrad.	テオシント	teosinte
Eupatorium chinense L. var. *simplicifolium* (Makino) Kitam.	ヒヨドリバナ	thoroughwort
Euphorbia helioscopia L.	トウダイグサ	devil's-milk, wortweed, sun spurge
Euphorbia hirta L.	シマニシキソウ	spurge, wolf's milk
Euphorbia maculata L.	オオニシキソウ	
Euphorbia pseudochamaesyce Fisch.	ニシキソウ	spurge, wolf's milk
Euphorbia supina Rafin.	コニシキソウ	milk-purslane
Fagus crenata Blume	ブナ	Siebold's beech
Fallopia convolvulus (L.) A. Love	ソバカズラ	
Festuca arundinacea Schreb.	トールフェスク, オニウシノケグサ	tall fescue
Festuca myuros (L.) C. Gmelin	ナギナタガヤ	rat's-tail
Festuca ovina L.	ウシノケグサ, シープフェスク	sheep's fescue
Festuca parvigluma Steud.	トボシガラ	fescue-grass
Festuca pratensis Huds.	メドウフェスク	meadow fescue
Festuca rubra L.	オオウシノケグサ, レッドフェスク, クリーピングフェスク	red fescue, creeping fescue
× *Festulolium* Braunii	フェストロリウム	
Fimbristylis ovata (Burm fil.) Kern	ヤリテンツキ	
Fimbristylis squarrosa Vahl	アゼテンツキ	
Galinsoga ciliata (Raf.) Blake.	ハキダメギク	
Galinsoga parviflora Cav.	コゴメギク	
Galium spurium L. var. *echinospermon* (Wallr.) Hayek	ヤエムグラ	bedstraw
Gentiana scabra Bunge var. *buergeri* (Miq.) Maxim.	リンドウ	gentian
Gentiana thunbergii (G. Don) Griseb.	ハルリンドウ	gentian
Gentiana zollingeri Fawcett	フデリンドウ	gentian

植物名（学名順）

Geranium thunbergii Sieb. et Zucc.	ゲンノショウコ	carnesbill
Glycine max (L.) Merr.	ダイズ	soy bean
Glycine soja Sieb. et Zucc.	ツルマメ	
Gnaphalium affine D. Don	ハハコグサ	cudweed, everlasting
Gnaphalium calviceps L.	タチチチコグサ	
Gnaphalium japonicum Thunb.	チチコグサ	cudweed, everlasting
Haloragis micrantha (Thunb.) R. Br.	アリノトウグサ	creeping raspwort
Helianthus annuus L.	ヒマワリ	sunflower
Hemarthria altissima (Poir.) Stapf et Hubbard	リンポグラス	limpograss
Hemarthria sibirica (Gandog.) Ohwi	ウシノシッペイ	
Hemerocallis fulva L. var. *kwanso* Regel	ヤブカンゾウ	day-lily
Hemerocallis longituba Miq.	ノカンゾウ	day-lily
Hemistepta lyrata Bunge	キツネアザミ	
Heteropogen contortus (L.) Beauv.	スペアグラス	speargrass
Hieracium aurantiacum L.	コウリンタンポポ	orange hawkweed
Hierochloe bungeana Trin.	コウボウ	
Holcus lanatus L.	シラゲガヤ，ベルベットグラス	velvetgrass
Hordeum vulgare L.	オオムギ	barley
Hosta albo-marginata (Hook.) Ohwi	コバギボウシ	narrow leaved plantain lily
Hydrangea macrophylla (Thunb.) Ser. f. *normalis* (Wilson) Hara	ガクアジサイ	hydrangea
Hydrangea macrophylla (Thunb.) Ser. var. *acuminata* (Sieb. et Zucc.) Makino	ヤマアジサイ	hydrangea
Hydrangea paniculata Siebold	ノリウツギ	hydrangea
Hydrocotyle maritima Honda	ノチドメ	water-pennywort, navelwort
Hydrocotyle ramiflora Maxim.	オオチドメ	water-pennywort, navelwort
Hydrocotyle sibthorpioides Lam.	チドメグサ	lawn pennywort, shining pennywort
Hypericum erectum Thunb.	オトギリソウ	St. John's-wort

Hypericum japonicum Thunb.	ヒメオトギリ	matted St. John's wort
Hypochoeris radicata L.	ブタナ	
Imperata cylindrica (L.) Beauv. var. *koenigii* (Retz.) Durand ét Schinz	チガヤ	needlegrass, cogongrass
Indigofera pseudo-tinctoria Matsum.	コマツナギ	indigo
Inula salicina L. var. *asiatica* Kitam.	カセンソウ	willow-leaved-inule
Ipomoea batatas (L.) Lam.	サツマイモ, カンショ（甘藷）	sweet potato
Ipomoea coccinea L.	マルバルコウ	scarlet morningglory
Ipomoea hederacea (L.) Jacq.	アメリカアサガオ	
Ipomoea hederacea (L.) Jacq. var *integriuscula* A. Gray	マルバアメリカアサガオ	
Ipomoea lacunosa L.	マメアサガオ	small-flowered white morning-glory
Ipomoea purpurea (L.) Roth	マルバアサガオ	
Ipomoea triloba L.	ホシアサガオ	
Iris ensata Thunb. var. *spontanea* (Makino) Nakai	ノハナショウブ	iris
Isachne globosa (Thunb.) O. Kuntze	チゴザサ	
Ixeris debilis (Thunb.) A. Gray	オオヂシバリ	
Ixeris dentata (Thunb.) Nakai	ニガナ	
Jacquemontia tamnifolia (L.) Griseb.	オキナアサガオ	
Juncus effusus L. var. *decipiens* Buchen.	イ	soft rush, mat rush, rush
Juncus tenuis Willd.	クサイ	path rush
Kalimeris pinnatifida (Maxim.) Kitam.	ユウガギク	
Kalimeris yomena Kitam.	ヨメナ	
Lactuca indica L. var. *laciniata* (O. Kuntze) Hara	アキノノゲシ	milkweed
Lamium amplexicaule L.	ホトケノザ	henbit, perfoliate-archangel henbit, deadnettle
Lamium purpureum L.	ヒメオドリコソウ	purple-dead-nettle, red deadnettle
Lepidium virginicum L.	マメグンバイナズナ	poor-man's pepper

Lespedeza cuneata (Du Mont. d. Cours.) G. Don	メドハギ	sericea lespedeza
Lespedeza cyrtobotrya Miq.	マルバハギ	bush-clover
Lespedeza pilosa (Thunb.) Sieb. et Zucc.	ネコハギ	bush-clover
Lespedeza striata (Thunb.) Schindler	ヤハズソウ	common lespedeza, Japan clover
Leucaena leucocephala (Lam.) De Wit.	ギンネム	leucaena
Leucosceptrum japonicum (Miq.) Kitam. et Murata	テンニンソウ	
Lilium auratum Lindl.	ヤマユリ	goldband lily, golden lily
Linaria canadensis Dum.	マツバウンラン	
Liriope minor (Maxim.) Makino	ヒメヤブラン	lily-turf
Lolium multiflorum Lam.	イタリアンライグラス, ネズミムギ	Italian ryegrass
Lolium perenne L.	ペレニアルライグラス, ホソムギ	perennial ryegrass
Lolium rigidum Gaud.	ウイメラライグラス	wymmera ryegrass
Lotus corniculatus L.	セイヨウミヤコグサ	common birdsfoot trefoil
Lotus corniculatus L. var. *japonicus* Regel	ミヤコグサ, バーズフットトレフォイル	birdsfoot trefoil
Ludwigia decurrens Walt.	ヒレタゴボウ	
Lupinus angustifolius L.	ルーピン (青花)	blue lupin
Lycopus lucidus Turcz.	シロネ	
Lysimachia clethroides Duby	オカトラノオ	loosestrife
Lysimachia japonica Thunb.	コナスビ	loosatrife
Macleaya cordata (Willd.) R. Br.	タケニグサ	plume poppy, tree celandine
Macroptilium atropurpureum (DC.) Urb.	サイラトロ	siratro
Macroptilium lathyroides (L.) Urb.	ファジービーン	phasey bean
Malus sieboldii (Regel) Rehder	ズミ	toringo crab
Malva neglecta Wallr.	ゼニバアオイ	
Malva parviflora L.	ウサギアオイ	
Matricaria chamomilla L.	カミツレ	
Matricaria matricarioides Port	コシカギク	
Mazus miquelii Makino	ムラサキサギゴケ	

Mazus pumilus (Burm. fil.) V. Steenis	トキワハゼ	
Medicago arabica (L.) Huds.	モンツキウマゴヤシ	
Medicago falcata L.	キバナアルファルファ	yellow alfalfa
Medicago lupulina L.	コメツブウマゴヤシ	black medick
Medicago media Pers.	雑色アルファルファ	variegated alfalfa
Medicago minima (L.) Bartal.	コウマゴヤシ	little bur clover
Medicago polymorpha L.	ウマゴヤシ, バークローバ	bur clover
Medicago sativa L.	アルファルファ, ルーサン, ムラサキウマゴ	alfalfa, lucerne
Melilotus alba Medic.	シロバナシナガワハギ, シロバナスィートクローバ	white sweetclover
Melilotus indica All.	コシナガワハギ	
Melilotus officinalis Pallas	シナガワハギ, キバナスィートクローバ	
Melinis minutiflora Beauv.	モラセスグラス	molasses grass
Melochia corchorifolia L.	ノジアオイ	
Miscanthus condensatus Hack.	ハチジョウススキ	eulalia
Miscanthus floridulus (Labill.) Warb.	トキワススキ	eulalia
Miscanthus oligostachyus Stapf	カリヤスモドキ	eulalia
Miscanthus sacchariflorus (Maxim.) Benth.	オギ	eulalia
Miscanthus sinensis Anderss.	ススキ	silvergrass, Japanese plume-grass
Miscanthus sinensis Anderss. f. *gracillimus* (Hitchc.) Ohwi	イトススキ	eulalia
Miscanthus tinctorius (Steud.) Hack.	カリヤス	eulalia
Moehringia lateriflora (L.) Fenzl	オオヤマフスマ	grove-sandwort
Muhlenbergia japonica Steud.	ネズミガヤ	muhly
Nicandra physaloides Gaertn.	オオセンナリ	
Oenothera biennis L.	メマツヨイグサ	common evening primrose
Oenothera erythrosepala Borbas	オオマツヨイグサ	evening primrose
Oenothera parviflora L.	アレチマツヨイグサ	evening primrose
Oenothera striata Ledeb.	マツヨイグサ	

Oplismenus undulatifolius (Ard.) Roemer et Schultes	ケチヂミザサ	
Oplismenus undulatifolius (Ard.) Roemer et Schultes var. *japonicus* (Steud.) Koidz.	チヂミザサ	
Oryza sativa L.	イネ	rice
Osmundastrum cinnamomeum (L.) Pr. var. *fokiense* (Copel) Tagawa	ヤマドリゼンマイ	
Oxalis corniculata L.	カタバミ	creeping wood sorrel, creeping lady's-sorrel
Paederia scandens (Lour.) Merrill var. *mairei* (Léveille) Hara	ヘクソカズラ	
Panicum antidotale Retz.	ブルーパニック	blue panic
Panicum bisulcatum Thunb.	ヌカキビ	panicum
Panicum coloratum L.	カラードギニアグラス	colored guineagrass
Panicum coloratum L. (Kabulabula type)	カブラブラグラス	kabulabulagrass
Panicum coloratum L. var. *makarikariense* Goossens	マカリカリグラス	makarikarigrass
Panicum dichotomiflorum Michx.	オオクサキビ	fall panicum
Panicum maximum Jacq.	ギニアグラス	guineagrass
Panicum maximum Jacq. var. *trichoglume* Eyles	グリーンパニック	green panic
Panicum miliaceum L.	キビ	millet, common millet
Panicum repens L.	ハイキビ	torpedograss
Panicum virgatum L.	スイッチグラス	switchgrass
Parnassia palustris L.	ウメバチソウ	grass-of-parnassus, white buttercups
Pascopyrum smithii (Rydb.) Löve	ウェスタンホィートグラス	western wheatgrass
Paspalum conjugatum Bergius	オガサワラスズメノヒエ	
Paspalum dilatatum Poir.	ダリスグラス, シマスズメノヒエ	dallisgrass, large watergrass
Paspalum distichum L.	キシュウスズメノヒエ	knotgrass, water couch
Paspalum notatum Flugge	バヒアグラス	bahiagrass
Paspalum thunbergii Kunth	スズメノヒエ	knotgrass

Paspalum urvillei Steud.	タチスズメノヒエ, ベージーグラス	vaseygrass
Patrinia scabiosaefolia Fisch.	オミナエシ	
Patrinia villosa (Thunb.) Juss.	オトコエシ	
Pennisetum alopecuroides (L.) Spreng.	チカラシバ	
Pennisetum clandestinum Hochst. ex Chiov.	キクユグラス	kikuyugrass
Pennisetum purpureum Schum.	ネピアグラス	napiergrass, elephantgrass
Pennisetum typhoides (Burmf.) Stapf et Hubb.	パールミレット, トウジンビエ	pearl millet
Petasites japonicus (Sieb. et Zucc.) Maxim.	フキ	lagwort
Phalaris arundinacea L.	リードカナリーグラス, クサヨシ	reed canarygrass
Phalaris tuberosa L. var. *stenoptera* (Hack.) Hitchc.	ハーディンググラス	hardinggrass
Phleum pratense L.	チモシー, オオアワガエリ	timothy
Phragmites communis Trin.	ヨシ	common reed
Phyllanthus urinaria L.	コミカンソウ	spurge
Physalis angulata L.	センナリホオズキ	
Physalis heterophylla Nees	アメリカホオズキ	
Pieris japonica (Thunb.) D.Don	アセビ	lily-of-the valley bush
Pinus densiflora Sieb. et Zucc.	アカマツ	Japanese red pine
Pinus thunbergii Parlat.	クロマツ	Japanese black pine
Plantago asiatica L.	オオバコ	plantain
Plantago lanceolata L.	ヘラオオバコ	buckthorn plantain, ribbed plantain, English plantain
Plantago major L.	セイヨウオオバコ	broad-leaved
Plantago virginica L.	ツボミオオバコ	white dwarf plantain, bearded plantain
Platycodon grandiflorum (Jacq.) A. DC.	キキョウ	ballon flower, Chinese bellflower, Japanese beltflower
Pleioblastus chino (Franch. et Savat.) Makino	アズマネザサ	
Pleioblastus chino (Franch. et Savat.) Makino	ネザサ	
Poa annua L.	スズメノカタビラ	annual bluegrass

植物名（学名順）

Poa compressa L.	コイチゴツナギ, カナダブルーグラス	Canada bluegrass
Poa nipponica Koidz.	オオイチゴツナギ	bluegrass, meadow-grass
Poa pratensis L.	ケンタッキーブルーグラス, ナガハグサ	Kentucky bluegrass
Poa sphondylodes Trin.	イチゴツナギ	bluegrass, meadowgrass
Poa trivialis L.	オオスズメノカタビラ	rough stalked meadowgrass
Polygonum aviculare L.	ミチヤナギ	blood-wort, bird's-knotgrass centynody, yard knotweed
Polygonum cuspidatum Sieb. et Zucc.	イタドリ	Japanese knotweed
Polygonum lapathifolium L.	オオイヌタデ	pale persicaria, willow smartweed, pale smartweed
Polygonum longisetum De Bruyn	イヌタデ	knotweed, smartweed
Polygonum persicaria L.	ハルタデ	persicary, red-legs, peachwort, lady's-thumb
Polygonum pilosum Roxb.	オオケタデ	
Polygonum sachalinense Fr. Schm.	オオイタドリ	sachaline giant knotweed
Populus sieboldii Miq.	ヤマナラシ	Japanese aspen
Potentilla chinensis Ser.	カワラサイコ	cinquefoil, potentil, five-finger
Potentilla fragarioides L. var. *major* Maxim.	キジムシロ	cinquefoil, potentil, five-finger
Potentilla freyniana Bornm.	ミツバツチグリ	cinquefoil, potentil, five-finger
Prunella vulgaris L. var. *lilacina* (Nakai) Nakai	ウツボグサ	selfheal
Pteridium aquilinum (L.) Kuhn var. *latiusculum* (Desv.) Und.	ワラビ	eastern bracken, brackenfern
Pueraria lobata (Willd.) Ohwi	クズ	kudzu-vine, kudzu
Pueraria phaseoloides (Roxb.) Benth.	熱帯クズ	tropical kudzu

植物名（学名順） [213]

Pulsatilla cernua (Thunb.) Spreng.	オキナグサ	pasque-flower
Quercus acutissima Carruth.	クヌギ	Japanese chestunt oak
Quercus dentata Thunb.	カシワ	daimyo oak
Quercus mongolica Fischer var. *grosseserrata* (Blume) Rehd. et Wils.	ミズナラ	oak
Quercus serrata Thunb.	コナラ	oak
Ranunculus japonicus Thunb.	ウマノアシガタ	buttercup, crowfoot
Ranunculus silerifolius Lev.	キツネノボタン	buttercup, crowfoot
Raphanus raphanistrum L.	セイヨウノダイコン	
Rhododendron japonicum (A. Gray) Suringer	レンゲツツジ	azalea
Rhododendron kaempferi Planch.	ヤマツツジ	azalea
Robinia pseudo-acacia L.	ハリエンジュ	black locust
Rorippa sylvestris Bess.	キレハイヌガラシ	
Rosa multiflora Thunb.	ノイバラ	polyantha rose
Rosa rugosa Thunb.	ハマナス	rose
Rosa wichuraiana Crep	テリハノイバラ	memorial rose
Rubia akane Nakai	アカネ	madder
Rubus crataegifolius Bunge	クマイチゴ	bramble
Rubus microphyllus L. fil.	ニガイチゴ	bramble
Rubus palmatus Thunb. var. *coptophyllus* (A. Gray) Koid.	モミジイチゴ	bramble
Rubus parvifolius L.	ナワシロイチゴ	bramble
Rudbeckia laciniata L.	オオハンゴンソウ	cut-leaf coneflower
Rumex acetosa L.	スイバ	sorrel
Rumex acetosella L.	ヒメスイバ	red sorrel
Rumex conglomeratus Murr.	アレチギシギシ	cluster dock
Rumex crispus L.	ナガバギシギシ	curly dock
Rumex japonicus Houtt.	ギシギシ	dock
Rumex obtusifolius L.	エゾノギシギシ	broadleaf dock
Saccharum officinarum L.	サトウキビ	sugar cane
Sagina japonica (Sw.) Ohwi	ツメクサ	pearlwort
Salix bakko Kimura	バッコヤナギ	willow
Salix integra Thunb.	イヌコリヤナギ	willow
Salix vulpina Anders.	キツネヤナギ	willow
Sanguisorba officinalis L.	ワレモコウ	burnet-blood wort
Sasa kurilensis (Rupr.) Makino et Shibata	チシマザサ	sasa, dwarf bamboo

植物名（学名順）

Sasa nipponica (Makino) Makino et Shibata	ミヤコザサ	sasa, dwarf bamboo
Sasa palmata (Marliac) Nakai	チマキザサ	sasa, dwarf bamboo
Sasa senanensis (Franch. et Savat.) Rehder	クマイザサ	sasa, dwarf bamboo
Sasa veitchii (Carr.) Rehder	クマザサ	sasa, dwarf bamboo
Sasamorpha borealis (Hack.) Nakai	スズタケ	
Scabiosa japonica Miq.	マツムシソウ	scabious
Scopolia japonica Maxim.	ハシリドコロ	
Secale cereale L.	ライムギ	rye
Senecio cannabifolius Less.	ハンゴンソウ	groundsel, ragwort, squaw-weed
Sesbania exaltata Cory et Rydb.	アメリカツノクサネム	
Setaria faberi Herrm.	アキノエノコログサ	foxtail grass, bristle grass
Setaria glauca (L.) Beauv.	キンエノコロ	yellow bristlegrass, yellow foxtailgrass, bottlegrass
Setaria italica Beauv.	アワ	foxtail millet, barngrass, Italian millet, Chinese corn
Setaria sphacelata (Schumach) Stapf et Hubb.	セタリア	golden timothy
Setaria viridis (L.) Beauv.	エノコログサ	green bristle-grass, green foxtail
Sicyos angulatus L.	アレチウリ	
Sida spinosa L.	アメリカキンゴジカ	
Siegesbeckia pubescens (Makino) Makino	メナモミ	
Silene conoidea L.	オオシラタマソウ	
Silene gallica L.	シロバナマンテマ	small-flowered catchfly, common catchfly
Silene gallica L. var. *quinquevulnera* (L.) Rohrb.	マンテマ	campion
Silybum marianum Gaertn.	オオアザミ	
Sisymbrium officinale Scop.	カキネガラシ	
Smilax china L.	サルトリイバラ	greenbrier, catbrier
Solanum aculeatissimum Jacq.	キンギンナスビ	
Solanum americanum Mill.	アメリカイヌホオズキ	

Solanum carolinense L.	ワルナスビ	horse nettle
Solanum nigrum L.	イヌホオズキ	hound-berry, fox-grape
Solidago altissima L.	セイタカアワダチソウ	tall goldenrod
Solidago virgaurea L. var. *asiatica* Nakai	アキノキリンソウ	goldenrod
Sonchus asper (L.) Hill	オニノゲシ	spiny sowthistle
Sonchus oleraceus L.	ノゲシ	sow-thistle
Sophora flavescens Solander ex Aiton var. *angustifolia* (Sieb. et Zucc.) Kitagawa	クララ	sophora
Sorghum almum Parodi.	コロンブスグラス	Columbus grass
Sorghum bicolor Moench	モロコシ, ソルガム	sorghum, sorgo
Sorghum halepense (L.) Pers.	ジョンソングラス, セイバンモロコシ	Johnsongrass
Sorghum sudanense (Piper) Stapf	スーダングラス	Sudangrass
Spergula arvensis L.	ノハラツメクサ	
Sphenoclea zeylanica Gaertn.	ナガボノウルシ	
Spiranthes sinensis (Pers.) Ames	ネジバナ	lady's-tresses, pearl-twist
Spodiopogon sibiricus Trin.	オオアブラススキ	
Sporobolus fertilis (Steud.) W. Clayton	ネズミノオ	smutgrass
Stellaria alsine Grimm var. *undulata* (Thunb.) Ohwi	ノミノフスマ	slender slarwort bog stitchwort
Stellaria aquatica (L.) Scop.	ウシハコベ	giant chickweed
Stellaria media (L.) Villars	ハコベ	chickweed
Stenotaphrum secundatum O. Kuntze.	セントオーガスチングラス	St. Augustine grass
Stylosanthes guianensis (Aubl.) Sw.	スタイロ	stylo
Stylosanthes humilis H. B. K.	タウンスビルスタイロ	townsville stylo, townsville lucerne
Swertia japonica (Schult.) Makino	センブリ	Columbo
Symphytum officinale L.	コンフリー	comfrey
Tagetes minuta L.	シオザキソウ	
Taraxacum albidum Dahlst.	シロバナタンポポ	dandelion, blowballs
Taraxacum hondoense Nakai ex H. Koidz.	エゾタンポポ	dandelion, blowballs
Taraxacum officinale Weber	セイヨウタンポポ	dandelion

Thalictrum minus L. var. *hypoleucum* (Sied. et Zucc.) Miq.	アキカラマツ	meadow-rue
Themeda japonica (Willd.) C. Tanaka	メガルカヤ	
Thesium chinense Turcz.	カナビキソウ	bastard-toad-flax
Thlaspi arvense L.	グンバイナズナ	
Tricyrtis affinis Makino	ヤマジノホトトギス	Japanese toad-lily
Trifolim campestre Schreb.	クスダマツメクサ	
Trifolium dubium Sibth.	コメツブツメクサ	yellow suckling clover, small hop clover
Trifolium fragiferum L.	ストロベリークローバ	strawberry clover
Trifolium hybridum L.	アルサイククローバ	alsike clover
Trifolium incarnatum L.	クリムソンクローバ	crimson clover
Trifolium pratense L.	アカクローバ, アカツメクサ	red clover
Trifolium repens L.	シロクローバ, シロツメクサ	white clover
Trifolium subterraneum L.	サブタレニアンクローバ	subterranean clover, subclover
Trigonotis peduncularis (Trevir.) Benth.	タビラコ	
Trisetum bifidum (Thunb.) Ohwi	カニツリグサ	
× *Triticosecale* Wittmack	ライコムギ	triticale
Triticum aestivum L.	コムギ	wheat
Vandellia crustacea (L.) Benth.	ウリクサ	
Veratrum maackii Regel var. *parviflorum* (Miq.) Hara et Mizushima	アオヤギソウ	false hellebore
Verbascum blattaria L.	モウズイカ	
Verbascum thapsus L.	ビロードモウズイカ	
Veronica arvensis L.	タチイヌノフグリ	corn speedwell, wall speedwell
Veronica hederaefolia L.	フラサバソウ	
Veronica persica Poir.	オオイヌノフグリ	birdseye, speedwell, cat's eyes
Viburnum dilatatum Thunb.	ガマズミ	viburnum, guelder-rose

Vicia cracca L.	クサフジ	bird vetch, tufted vetch, gerard vetch
Vicia hirsuta (L.) S.F. Gray	スズメノエンドウ	common hairy tare, tyne grass
Vicia sepium L.	コモンベッチ, カラスノエンドウ	common vetch
Vicia tetrasperma (L.) Schreb.	カスマグサ	lentillare, smooth tare
Vicia villosa Roth	ヘアリーベッチ	hairy vetch
Vigna sinensis (L.) Savi ex Hassk.	カウピー	cowpea
Viola grypoceras A. Gray	タチツボスミレ	violet pansy
Weigela hortensis (Sieb. et Zucc.) K. Koch	タニウツギ	
Wistaria floribunda (Willd.) DC.	フジ	Japanese wistaria
Xanthium occidentale Bertoloni	オオオナモミ	
Xanthium strumarium L.	オナモミ	cocklebur
Youngia denticulata (Houtt.) Kitam.	ヤクシソウ	
Youngia japonica (L.) DC.	オニタビラコ	oriental hawkbeard
Zea mays L.	トウモロコシ	corn, maize
Zoysia japonica Steud.	シバ	Japanese lawngrass
Zoysia macrostachya Franch. et Savat.	オニシバ	
Zoysia matrella (L.) Merr.	コウシュンシバ	Manilagrass, Manila-bluegrass
Zoysia tenuifolia Willd.	コウライシバ	Korean lawn-grass, Koreangrass, mascarenegrass

病害名 （和名順）

〈対象作物略号〉
イ共：イネ科牧草共通，オ：オーチャードグラス，チ：チモシー，
バ：バーミューダグラス，フ：フェスク，ブ：ブロムグラス，
ラ：ライグラス，リ：リードカナリーグラス，レ：レッドトップ，
マ共：マメ科共通，ア：アルファルファ，ク：クローバ，
エ：エンバク，ソ：ソルガム，ト：トウモロコシ

和 名	対象作物	学 名	英 名
赤かび病	ト，エ，イ共	Gibberella zeae (Schwein.) Petch, Fusarium avenaceum (Fr.) Sacc.	Scab, Fusarium blight
赤さび病	リ	Puccinia sessilis Schneid.: Schröt. var. sessilis	Leaf rust
網斑病	イ共	Drechslera dictyoides (Drechs.) Shoem.	Net blotch
萎凋病	マ共	Fusarium oxysporum Schlechtend.: Fr. f.sp. medicaginis (Weimer) Snyder & Hans.	Wilt
いぼ斑点病	マ共	Pseudopeziza spp.	Pseudopeziza leaf spot, Common leaf spot
いもち病	ト，エ，イ共	Magnaporthe grisea (Hebert) Yaegashi & Udagawa	Blast
うどんこ病	イ共	Blumeria graminis (DC.) Speer	Powdery mildew
うどんこ病	マ共	Microsphaeria spp., Erysiphe pisi DC.	Powdery mildew
黄化萎縮病	ト，ソ，イ共	Sclerophthora macrospora (Sacc.) Thirum., Shaw et Naras.	Downy mildew
黄斑病	ア	Sporonema phacidioides Desmaz.	Yellow leaf blotch
汚斑病	ク	Curvularia trifolii (Kauffm.) Boedijn	Curvularia leaf blight
角斑病	イ共	Pseudoseptoria stomaticola (Bäumler) Sutton	Stem speckle

病害名（和名順）

和名	略号	学名	英名
かさ枯病	イ共	*Pseudomonas syringae* pv. *atropurpurea* (Reddy & Godkin 1923) Young, Dye & Wilkie 1978	Halo blight
褐条病	ト，イ共	*Acidovorax avenae* subsp. *avenae* (Manns 1909) Willems, Goor, Thielemans, Gillis, Kersters & De Ley 1992	Brown stripe
褐条べと病	ト	*Sclerophthora rayssiae* Kenneth, Kaltin & Wahl var. *zeae* Payak & Renfro	Brown stripe downy mildew
褐色雪腐病	イ共，マ共	*Pythium iwayamai* Ito, *P. paddicum* Hirane, *Pythium* spp.	Pythium snow blight
褐斑病	ト	*Kabatiella zeae* Narita & Hiratsuka	Eye spot
褐斑病	リ	*Stagonospora foliicola* (Bres.) Bubák	Buff spot
褐斑病	ク	*Stagonospora recedens* (Massal.) Jones & Weimer	Leaf spot
株腐病	ラ	*Rhizoctonia cerealis* Van der Hoeven	Foot rot
がまの穂病	イ共	*Epichloë typhina* (Pers.: Fr.) Tul. [*Neotyphodium typhinum* (Morgan-Jones & Gams) Glenn, Bacon & Hanlin]	Choke
冠さび病	イ共	*Puccinia coronata* Corda var. *coronata*	Crown rust
黄さび病	イ共	*Puccinia striiformis* Westend.	Stripe rust
菌核病	マ共	*Sclerotinia trifoliorum* Eriks.	Sclerotinia crown and stem rot
茎枯病	ア	*Phoma medicaginis* Malbr. & Roum. var. *medicaginis* Boerema	Spring black stem
茎割病	ク	*Kabatiella caulivora* (Kirchn.) Karak.	Northern anthracnose
雲形病	オ	*Rhynchosporium orthosporum* Caldwell	Scald

雲形病	イ共	*Rhynchosporium secalis* (Oudem.) Davis	Scald
黒あし病	ア	*Cylindrocladium crotalariae* (Loos) Bell & Sobers, *C. floridanum* Sobers & Seymour	Crown rot
黒かび病	ク	*Rhizoctonia leguminicola* Gough & Elliott	Blackpatch
黒ごま病	イ共	*Phyllachora graminis* (Pers.: Fr.) Nitschke ex Fuckel	Tar spot
黒さび病	イ共	*Puccinia graminis* Pers.: Pers. subsp. *graminicola* Urban	Stem rust
黒穂病	イ共	*Ustilago* spp.	Smut
黒穂病	ト	*Ustilago maydis* (DC.) Corda	Smut
こうがいかび病	ク	*Choanephora cucurbitarum* (Berk. & Ravenel) Thaxt.	Choanephora rot
紅色雪腐病	イ共	*Monographella nivalis* (Schaffnit) Müller	Pink snow mold
黒斑病	レ	*Cheilaria agrostis* Lib.	Blotch and char spot
小さび病	オ	*Uromyces dactylidis* Otth var. *dactylidis*	Uromyces rust
ごま葉枯病	ト	*Cochliobolus heterostrophus* (Drechs.) Drechs.	Soutern leaf blight
さび病	ト	*Puccinia sorghi* Schwein.	Common rust
さび病	マ共	*Uromyces* spp.	Rust
シスト線虫病	マ共	*Heterodera trifolii* Goffart	Cyst
縞葉枯病	ト	Rice stripe virus (RSV)	Stripe
蛇紋病	ラ	*Ascochyta desmazieresii* Cavara.	Leaf blotch
条斑細菌病	ト, ソ	*Burkholderia andropogonis* (Smith 1911) Gillis, Van Van, Bardin, Goor, Hebbar, Willems, Segers, Kersters, Heulin & Fernandez 1995	Bacterial stripe
白絹病	ト, ソ, イ共, マ共	*Sclerotium rolfsii* Sacc. [*Athelia rolfsii* (Curzi) Tu & Kimbrough]	Southern blight, Sclerotium blight
紫輪病	ソ	*Cercospora sorghi* Ellis & Everhart	Gray leaf spot

[222] 病害名（和名順）

和名		学名	英名
白枯病	バ	*Cochliobolus cynodontis* Nelson	Leaf blight
すじ萎縮病	ト	Rice black streaked dwarf virus (RBsDV)	Streaked dwarf
すじ黒穂病	イ共	*Ustilago striiformis* (Westend.) Niessl	Stripe smut
すじ葉枯病	イ共	*Cercosporidium graminis* (Fuckel) Deighton	Brown stripe, Leaf streak
すす点病	ク	*Cymadothea trifolii* (Killian) Wolf	Sooty blotch, Black blotch
すす紋病	ト, ソ	*Exserohilum turcicum* (Pass.) Leonard & Suggs	Northern leaf blight
炭腐病	マ共	*Macrophomina phaseolina* (Tassi) Goidánich	Charcoal rot
そばかす病	マ共	*Leptosphaerulina trifolii* (Rostr.) Petr.	Leptosphaerulina leaf spot, Pepper spot
立枯病	イ共	*Gaeumannomyces graminis* (Sacc.) Arx & Olivier	Take-all
炭疽病	ト, ソ, エ, イ共	*Colletotrichum graminicola* (Ces.) Wils.	Anthracnose
炭疽病	マ共	*Colletotrichum* spp.	Southern anthracnose, Anthracnose
てんぐ巣病	ク	Phytoplasma	Witches' broom
鳥の目病	チ	*Ovularia pusilla* (Unger) Sacc.	Bird's-eye spot
苗立枯病	ト	*Fusarium avenaceum* (Fries) Saccardo, *Penicillium* sp.	Seedling blight, Damping-off
夏葉枯病	オ	*Drechslera dactylidis* Shoem.	Drechslera leaf blight
夏斑点病	ラ	*Cochliobolus sativus* (Ito & Kuribayashi) Drechs. ex Dastur	Helminthosporium leaf spot
なまぐさ黒穂病	リ	*Tilletia sphaerococca* (Wallr.) Fisch. v.	Bunt
南方さび病	ト	*Puccinia polysora* Underw.	Southern rust
根腐病	ト	*Pythium graminicola* Subram.	Browning root rot
根腐線虫病	イ共, マ共	*Pratylenchus coffeae* (Zimmermann) Filipjev & Stekhoven, *P. penetrans* (Cobb) Filipjev & Stekhoven	Root lesion
根朽病	ト	*Rhizoctonia solani* Kühn AG2-2	Root rot

病害名（和名順）　[223]

和名	寄主	学名	英名
根こぶ線虫病	イ共, マ共	*Meloidogyne arenaria* (Neal) Chitwood, *M. hapla* Chitwood, *M. incognita* (Kofoid & White) Chitwood, *M. javanica* (Treub) Chitwood	Root knot
ねん葉病	オ	*Dilophospora alopecuri* (Fr.) Fr.	Twist
バーティシリウム萎凋病	ア	*Verticillium albo-atrum* Reinke & Berthold	Verticillium wilt
灰色かび病	イ共, マ共	*Botrytis cinerea* Pers.: Fr.	Gray mold
葉枯病	チ	*Drechslera phlei* (Graham) Shoem.	Leaf blight, Leaf blotch
葉枯病	オ	*Stagonospora arenaria* Sacc.	Leaf blotch
葉枯病	フ	*Septoria tenella* Cooke & Ellis	Leaf spot
葉枯病	マ共	*Pleospora herbarum* (Pers.: Fr.) Rabenh.	Stemphylium leaf blight
葉腐病	イ共, マ共	*Rhizoctonia solani* Kühn AG1 I A, I B	Summer blight, Rhizoctonia rot
白斑病	ク	*Leptosphaeria pratensis* Saccardo & Briard	Stagonospora leaf spot
麦角病	イ共	*Claviceps purpurea* (Fr.:Fr.) Tul.	Ergot
麦角病	ソ	*Claviceps sorghicola* Tsukiboshi, Shimanuki & Uematsu, *Sphacelia sorghi* Mcrae	Ergot
斑点病	ト	*Physoderma maydis* Miyabe	Brown spot, Brown bundle disease
斑点病	オ	*Mastigosporium rubricosum* (Dearn. & Barth.) Nannf.	Eye spot
斑点病	ラ	*Drechslera siccans* (Drechs.) Shoem.	Leaf blight
斑点病	フ, ブ	*Cochliobolus sativus* (Ito et Kuribayashi) Drechs. ex Dastur [*Drechslera siccans* (Drechs.) Shoem.]	Leaf spot
斑点病	チ	*Cladosporium phlei* (Gregory) De Vries	Purple spot
斑点病	マ共	*Cercospora zebrina* Pass.	Summer black stem and leaf spot

病害名（和名順）

和名		学名	英名
斑点細菌病	ク	*Pseudomonas syringae* pv. *syringae* van Hall 1902	Bacterial leaf spot
ピシウム苗立枯病	ト	*Pythium* spp.	
ひょう紋病	ト，ソ，イ共	*Gloeocercospora sorghi* Bain & Edgerton ex Deighton	Zonate leaf spot, Copper spot
フィトフトラ根腐病	ア	*Phytophthora medicaginis* Hansen & Maxwell	Phytophthora root rot
縁枯病	イ共	*Drechslera nobleae* Mckenzie & Matthews	Drechslera blight
べと病	マ共	*Peronospora trifoliorum* de Bary	Downy mildew
北方斑点病	ト	*Cochliobolus carbonum* Nelson	Northern leaf spot
ミイラ穂病	イ共	*Ephelis* sp.	Black choke
紫斑点病	オ	*Stagonospora maculata* (Grove) Sprague	Purple leaf spo
紫斑点病	ソ	*Bipolaris sorghicola* (Lefebvre & Sherwin) Alcorn	Target spot
紫紋羽病	マ共	*Helicobasidium mompa* Tanaka	Violet root rot
モザイク病	オ	*Cocksfoot mottle virus* (CfMV)	Mosaic
モザイク病	ト，ソ	*Sugarcane mosaic virus* (ScMV), *Cucumber mosaic virus* (CMV)	Mosaic
モザイク病	マ共	*Alfalfa mosaic virus* (AMV), *Bean yellow mosaic virus* (BYMV), *White clover mosaic virus* (WCMV)	Mosaic
紋枯病	ト，ソ	*Rhizoctonia solani* Kühn AG1 I A	Sheath blight
雪腐大粒菌核病	イ共	*Myriosclerotinia borealis* (Bubák & Vleugel) Kohn	Sclerotinia snow blight, Snow mold
雪腐褐色小粒菌核病	イ共，マ共	*Typhula incarnata* Lasch: Fr.	Typhula snow blight
雪腐黒色小粒菌核病	イ共，マ共	*Typhula ishikariensis* Imai var. *ishikariensis*	Typhula snow blight
輪紋病	マ共	*Stemphylium* spp.	Stemphylium leaf spot

病害名 (学名順)

〈対象作物略号〉
イ共:イネ科牧草共通, オ:オーチャードグラス, チ:チモシー,
バ:バーミューダグラス, フ:フェスク, ブ:ブロムグラス,
ラ:ライグラス, リ:リードカナリーグラス, レ:レッドトップ,
マ共:マメ科共通, ア:アルファルファ, ク:クローバ,
エ:エンバク, ソ:ソルガム, ト:トウモロコシ

学 名	和 名	対象作物	英 名
Acidovorax avenae subsp. *avenae* (Manns 1909) Willems, Goor, Thielemans, Gillis, Kersters & De Ley 1992	褐条病	ト, イ共	Brown stripe
Alfalfa mosaic virus (AMV), *Bean yellow mosaic virus* (BYMV), *White clover mosaic virus* (WCMV)	モザイク病	マ共	Mosaic
Ascochyta desmazieresii Cavara.	蛇紋病	ラ	Leaf blotch
Bipolaris sorghicola (Lefebvre & Sherwin) Alcorn	紫斑点病	ソ	Target spot
Blumeria graminis (DC.) Speer	うどんこ病	イ共	Powdery mildew
Botrytis cinerea Pers.: Fr.	灰色かび病	イ共, マ共	Gray mold
Burkholderia andropogonis (Smith 1911) Gillis, Van Van, Bardin, Goor, Hebbar, Willems, Segers, Kersters, Heulin & Fernandez 1995	条斑細菌病	ト, ソ	Bacterial stripe
Cercospora sorghi Ellis & Everhart	紫輪病	ソ	Gray leaf spot
Cercospora zebrina Pass.	斑点病	マ共	Summer black stem and leaf spot
Cercosporidium graminis (Fuckel) Deighton	すじ葉枯病	イ共	Brown stripe, Leaf streak
Cheilaria agrostis Lib.	黒斑病	レ	Blotch and char spot
Choanephora cucurbitarum (Berk. & Ravenel) Thaxt.	こうがいかび病	ク	Choanephora rot
Cladosporium phlei (Gregory) De Vries	斑点病	チ	Purple spot

[226] 病害名（学名順）

Claviceps purpurea (Fr.: Fr.) Tul.	麦角病	イ共	Ergot
Claviceps sorghicola Tsukiboshi, Shimanuki & Uematsu, Sphacelia *sorghi* Mcrae	麦角病	ソ	Ergot
Cochliobolus carbonum Nelson	北方斑点病	ト	Northern leaf spot
Cochliobolus cynodontis Nelson	白枯病	バ	Leaf blight
Cochliobolus heterostrophus (Drechs.) Drechs.	ごま葉枯病	ト	Soutern leaf blight
Cochliobolus sativus (Ito & Kuribayashi) Drechs. ex Dastur	夏斑点病	ラ	Helminthosporium leaf spot
Cochliobolus sativus (Ito et Kuribayashi) Drechs. ex Dastur [*Drechslera siccans* (Drechs.) Shoem.]	斑点病	フ, ブ	Leaf spot
Cocksfoot mottle virus (CfMV)	モザイク病	オ	Mosaic
Colletotrichum graminicola (Ces.) Wils.	炭疽病	ト, ソ, エ, イ共	Anthracnose
Colletotrichum spp.	炭疽病	マ共	Southern anthracnose, Anthracnose
Curvularia trifolii (Kauffm.) Boedijn	汚斑病	ク	Curvularia leaf blight
Cylindrocladium crotalariae (Loos) Bell & Sobers, *C. floridanum* Sobers & Seymour	黒あし病	ア	Crown rot
Cymadothea trifolii (Killian) Wolf	すす点病	ク	Sooty blotch, Black blotch
Dilophospora alopecuri (Fr.) Fr.	ねん葉病	オ	Twist
Drechslera dactylidis Shoem.	夏葉枯病	オ	Drechslera leaf blight
Drechslera dictyoides (Drechs.) Shoem.	網斑病	イ共	Net blotch
Drechslera nobleae Mckenzie & Matthews	縁枯病	イ共	Drechslera blight
Drechslera phlei (Graham) Shoem.	葉枯病	チ	Leaf blight, Leaf blotch

Drechslera siccans (Drechs.) Shoem.	斑点病	ラ	Leaf blight
Ephelis sp.	ミイラ穂病	イ共	Black choke
Epichloë typhina (Pers.: Fr.) Tul. [*Neotyphodium typhinum* (Morgan-Jones & Gams) Glenn, Bacon & Hanlin]	がまの穂病	イ共	Choke
Exserohilum turcicum (Pass.) Leonard & Suggs	すす紋病	ト, ソ	Northern leaf blight
Fusarium avenaceum (Fries) Saccardo, *Penicillium* sp.	苗立枯病	ト	Seedling blight, Damping-off
Fusarium oxysporum Schlechtend.: Fr. f.sp. *medicaginis* (Weimer) Snyder & Hans.	萎凋病	マ共	Wilt
Gaeumannomyces graminis (Sacc.) Arx & Olivier	立枯病	イ共	Take-all
Gibberella zeae (Schwein.) Petch, *Fusarium avenaceum* (Fr.) Sacc.	赤かび病	ト, エ, イ	Scab, Fusarium blight
Gloeocercospora sorghi Bain & Edgerton ex Deighton	ひょう紋病	ト, ソ, イ共	Zonate leaf spot, Copper spot
Helicobasidium mompa Tanaka	紫紋羽病	マ共	Violet root rot
Heterodera trifolii Goffart	シスト線虫病	マ共	Cyst
Kabatiella caulivora (Kirchn.) Karak.	茎割病	ク	Northern anthracnose
Kabatiella zeae Narita & Hiratsuka	褐斑病	ト	Eye spot
Leptosphaeria pratensis Saccardo & Briardt	白斑病	ク	Stagonospora leaf spo
Leptosphaerulina trifolii (Rostr.) Petr.	そばかす病	マ共	Leptosphaerulina leaf spot, Peppe r spot
Macrophomina phaseolina (Tassi) Goidánich	炭腐病	マ共	Charcoal rot
Magnaporthe grisea (Hebert) Yaegashi & Udagawa	いもち病	ト, エ, イ	Blast
Mastigosporium rubricosum (Dearn. & Barth.) Nannf.	斑点病	オ	Eye spot

学名	和名	区分	英名
Meloidogyne arenaria (Neal) Chitwood, *M. hapla* Chitwood, *M. incognita* (Kofoid & White) Chitwood, *M. javanica* (Treub) Chitwood	根こぶ線虫病	イ共, マ共	Root not
Microsphaeria spp., *Erysiphe pisi* DC.	うどんこ病	マ共	Powdery mildew
Monographella nivalis (Schaffnit) Müller	紅色雪腐病	イ共	Pink snow mold
Myrioscleotinia borealis (Bubák & Vleugel) Kohn	雪腐大粒菌核病	イ共	Sclerotinia snow blight, Snow mold
Ovularia pusilla (Unger) Sacc.	鳥の目病	チ	Bird's-eye spot
Peronospora trifoliorum de Bary	べと病	マ共	Downy mildew
Phoma medicaginis Malbr. & Roum. var. *medicaginis* Boerema	茎枯病	ア	Spring black stem
Phyllachora graminis (Pers.: Fr.) Nitschke ex Fuckel	黒ごま病	イ共	Tar spot
Physoderma maydis Miyabe	斑点病	ト	Brown spot, Brown bundle disease
Phytophthora medicaginis Hansen & Maxwell	フィトフトラ根腐病	ア	Phytophthora root rot
Phytoplasma	てんぐ巣病	ク	Witches' broom
Pleospora herbarum (Pers.: Fr.) Rabenh.	葉枯病	マ共	Stemphylium leaf blight
Pratylenchus coffeae (Zimmermann) Filipjev & Stekhoven, *P. penetrans* (Cobb) Filipjev & Stekhoven	根腐線虫病	イ共, マ共	Root lesion
Pseudomonas syringae pv. *atropurpurea* (Reddy & Godkin 1923) Young, Dye & Wilkie 1978	かさ枯病	イ共	Halo blight
Pseudomonas syringae pv. *syringae* van Hall 1902	斑点細菌病	ク	Bacterial leaf spot
Pseudopeziza spp.	いぼ斑点病	マ共	Pseudopeziza leaf spot, Common leaf spot

Pseudoseptoria stomaticola (Bäumler) Sutton	角斑病	イ共	Stem speckle
Puccinia coronata Corda var. *coronata*	冠さび病	イ共	Crown rust
Puccinia graminis Pers.: Pers. subsp. *graminicola* Urban	黒さび病	イ共	Stem rust
Puccinia polysora Underw.	南方さび病	ト	Southern rust
Puccinia sessilis Schneid.: Schröt. var. *sessilis*	赤さび病	リ	Leaf rust
Puccinia sorghi Schwein.	さび病	ト	Common rust
Puccinia striiformis Westend.	黄さび病	イ共	Stripe rust
Pythium graminicola Subram.	根腐病	ト	Browning root rot
Pythium iwayamai Ito, *P. paddicum* Hirane, *Pythium* spp.	褐色雪腐病	イ共, マ共	Pythium snow blight
Pythium spp.	ピシウム苗立枯病	ト	
Rhizoctonia cerealis Van der Hoeven	株腐病	ラ	Foot rot
Rhizoctonia leguminicola Gough & Elliott	黒かび病	ク	Blackpatch
Rhizoctonia solani Kühn AG1 I A	紋枯病	ト, ソ	Sheath blight
Rhizoctonia solani Kühn AG1 I A, I B	葉腐病	イ共, マ共	Summer blight, Rhizoctonia rot
Rhizoctonia solani Kühn AG2-2	根朽病	ト	Root rot
Rhynchosporium orthosporum Caldwell	雲形病	オ	Scald
Rhynchosporium secalis (Oudem.) Davis	雲形病	イ共	Scald
Rice black streaked dwarf virus (RBsDV)	すじ萎縮病	ト	Streaked dwarf
Rice stripe virus (RSV)	縞葉枯病	ト	Stripe
Sclerophthora macrospora (Sacc.) Thirum., Shaw et Naras.	黄化萎縮病	ト, ソ, イ	Downy mildew
Sclerophthora rayssiae Kenneth, Kaltin & Wahl var. *zeae* Payak & Renfro	褐条べと病	ト	Brown stripe downy mildew
Sclerotinia trifoliorum Eriks.	菌核病	マ共	Sclerotinia crown and stem rot

学名	和名	作物	英名
Sclerotium rolfsii Sacc. [*Athelia rolfsii* (Curzi) Tu & Kimbrough]	白絹病	ト, ソ, イ共, マ共	Southern blight, Sclerotium blight
Septoria tenella Cooke & Ellis	葉枯病	フ	Leaf spot
Sporonema phacidioides Desmaz.	黄斑病	ア	Yellow leaf blotch
Stagonospora arenaria Sacc.	葉枯病	オ	Leaf blotch
Stagonospora foliicola (Bres.) Bubák	褐斑病	リ	Buff spot
Stagonospora maculata (Grove) Sprague	紫斑点病	オ	Purple leaf spot
Stagonospora recedens (Massal.) Jones & Weimer	褐斑病	ク	Leaf spot
Stemphylium spp.	輪紋病	マ共	Stemphylium leaf spot
Sugarcane mosaic virus (ScMV), *Cucumber mosaic virus* (CMV)	モザイク病	ト, ソ	Mosaic
Tilletia sphaerococca (Wallr.) Fisch. v.	なまぐさ黒穂病	リ	Bunt
Typhula incarnata Lasch: Fr.	雪腐褐色小粒菌核病	イ共, マ共	Typhula snow blight
Typhula ishikariensis Imai var. *ishikariensis*	雪腐黒色小粒菌核病	イ共, マ共	Typhula snow blight
Uromyces dactylidis Otth var. *dactylidis*	小さび病	オ	Uromyces rust
Uromyces spp.	さび病	マ共	Rust
Ustilago maydis (DC.) Corda	黒穂病	ト	Smut
Ustilago striiformis (Westend.) Niessl	すじ黒穂病	イ共	Stripe smut
Ustilago spp.	黒穂病	イ共	Smut
Verticillium albo-atrum Reinke & Berthold	バーティシリウム萎凋病	ア	Verticillium wilt

昆虫および動物名 （和名順）

和名	学名	英名
アオズキンヨコバイ	*Batracomorphus mundus* (Matsumura)	
アオバネサルハムシ (キアシサルハムシ)	*Basilepta fulvipes* (Motschulsky)	
アカカスリヨコバイ	*Balclutha rubinervis* (Matsumura)	
アカスジメクラガメ	*Stenotus rubrovittatus* (Matsumura)	sorghum plant bug
アカヒゲホソミドリメクラガメ	*Trigonotylus coelestialium* (Kirkaldy)	rice leaf bug
アカヒメヘリカメムシ	*Aeschyteles maculatus* (Fieber)	carrot bug
アカビロウドコガネ	*Maladera castanea* (Arrow)	Asiatic garden beetle
アカホシメクラガメ	*Creontiades pallidifer* (Walker)	
アシナガコガネ	*Hoplia communis* Waterhouse	
アズマモグラ	*Mogera wogura* (Temminck)	Japanese mole
アマヒトリ	*Phragmatobia fuliginosa japonica* Rothschild	flax arctid, ruby tiger
アヤモクメキリガ	*Xylena fumosa* (Butler)	rape caterpillar
アルファルファタコゾウムシ	*Hypera postica* (Gyllenhal)	alfalfa weevil
アレナリアネコブセンチュウ	*Meloidogyne arenaria* (Neal) Chitwood	peanut root-knot nematode
アワノメイガ	*Ostrinia furnacalis* (Guenée)	oriental corn borer
アワヨトウ	*Mythimna separata* (Walker)	oriental armyworm, rice armyworm, rice ear-cutting caterpillar
イチモンジカメムシ	*Piezodorus hybneri* (Gmelin)	
イナゴモドキ	*Parapleurus alliaceus* (Germar)	
イネクダアザミウマ	*Haplothrips aculeatus* (Fabricius)	rice aculeated thrips
イネマダラヨコバイ	*Recilia oryzae* (Matsumura)	
イネミギワバエ (イネヒメハモグリバエ)	*Hydrellia griseola* (Fallén)	rice leafminer
イネヨトウ	*Sesamia inferens* (Walker)	pink borer
ウコンノメイガ	*Pleuroptya ruralis* (Scopoli)	bean webworm
ウスイロガガンボ	*Tipula subcunctans* Alexander	
ウスイロササキリ	*Conocephalus chinensis* (Redtenbacher)	

昆虫および動物名（和名順）

和名	学名	英名
ウスカワマイマイ	*Acusta despecta sieboldiana* (Pfeiffer)	
ウスバミドリヨコバイ	*Balclutha viridis* (Matsumura)	
ウスミドリメクラガメ（ツマグロアオメクラガメ）	*Apolygus spinolai* (Meyer-Dür)	
ウリハムシ	*Aulacophora femoralis* (Motschulsky)	cucurbit leaf beetle
ウリハムシモドキ	*Atrachya menestriesi* (Faldermann)	false melon beetle
エゾスキガタキモグリバエ	*Cetema cereris* (Fallén)	
エゾホソガガンボ	*Nephrotoma cornicina* (Linnaeus)	
エゾユキウサギ	*Lepus timidus ainu* Barrett-Hamilton	Hokkaido hare
エンドウヒゲナガアブラムシ	*Acyrthosiphon pisum* (Harris)	pea aphid
エンマコオロギ	*Teleogryllus emma* (Ohmachi et Matsuura)	field cricket
オオタコゾウムシ	*Hypera punctata* (Fabricius)	clover leaf weevil
オオトガリヨコバイ	*Doratulina grandis* (Matsumura)	
オオトゲシラホシカメムシ	*Eysarcoris lewisi* (Distant)	
オオヨコバイ	*Cicadella viridis* (Linnaeus)	
オカダンゴムシ	*Armadillidium vulgare* Latreille	pillbug
オカモノアラガイ	*Succinea lauta* Gould	
オナガササキリ	*Conocephalus gladiatus* (Redtenbacher)	
カスリヨコバイ	*Balclutha punctata* (Fabricius)	
カバイロコメツキ	*Ectinus sericeus* (Candeze)	wheat wireworm
カブラヤガ	*Agrotis segetum* (Denis et Schiffermüller)	cutworm, turnip moth
カホンカハナアザミウマ	*Frankliniella tenuicornius* (Uzel)	
カラフトアオホソガガンボ	*Nephrotoma aculeata* (Loew)	
カンザワハダニ	*Tetranychus kanzawai* Kishida	Kanzawa spider mite
キイロホソガガンボ（キスジガガンボ）	*Nephrotoma virgata* (Coquillett)	
キジ	*Phasianus colchicus* Linnaeus	common pheasant
キジバト	*Streptopelia orientalis* (Latham)	rufous turtle dove

キタネグサレセンチュウ	*Pratylenchus penetrans* (Cobb) Filipjev et Schuurmans Stekhoven	Cobb root-lesion nematode
キタネコブセンチュウ	*Meloidogyne hapla* Chitwood	northern root-knot nematode
キハラゴマダラヒトリ	*Spilosoma lubricipedum* (Linnaeus)	
キボシアツバ	*Paragabara flavomacula* (Oberthür)	
キボシマルトビムシ	*Bourletiella hortensis* (Fitch)	garden springtail
キマダラヒロヨコバイ	*Scleroracus flavopictus* (Ishihara)	
キマルトビムシ	*Sminthurus viridis* (Linnaeus)	lucerne flea
ギンムジハマキ	*Eana (Ablabia) argentana* (Clerck)	
クサキイロアザミウマ	*Anaphothrips obscurus* (Müller)	grass thrips
クサシロキヨトウ	*Acantholeucania loreyi* (Duponchel)	
クシコメツキ	*Melanotus legatus* Candeze	
クビキリギリス	*Euconocephalus thunbergi* (Stål)	
クモヘリカメムシ	*Leptocorisa chinensis* (Dallas)	rice bug
クルマバッタモドキ	*Oedaleus infernalis* de Saussure	
クルミネグサレセンチュウ	*Pratylenchus vulnus* Allen et Jensen	walnut root-lesion nematode
クローバシストセンチュウ	*Heterodera trifolii* Goffart	clover cyst nematode
クローバタネコバチ	*Bruchophagus gibbus* (Boheman)	clover seed chalcid
クローバハダニ	*Bryobia praetiosa* koch	clover mite
クロクシコメツキ	*Melanotus senilis* Candeze	
クロコガネ	*Holotrichia kiotonensis* Brenske	balck chafer
クロフツノウンカ	*Perkinsiella saccharicida* Kirkaldy	sugarcane planthopper, sugarcane leafhopper
クロミャクイチモンジヨコバイ	*Exitianus indicus* (Distant)	
クワキヨコバイ	*Pagaronia guttigera* (Uhler)	yellow mulberry leafhopper

昆虫および動物名（和名順）

ケチビコフキゾウムシ	*Sitona hispidulus* (Fabricius)	clover root curculio
コアオハナムグリ	*Oxycetonia jucunda* (Faldermann)	citrus flower chafer
コウモリガ	*Endoclyta excrescens* (Butler)	swift moth
コガタクシコメツキ	*Melanotus erythropygus* Candeze	
コナガ	*Plutella xylostella* (Linnaeus)	diamondback moth, cabbage moth
コバネイナゴ	*Oxya yezoensis* Shiraki	rice grasshopper
コバネササキリ	*Conocephalus japonicus* (Redtenbacher)	
コフサキバガ	*Dichomeris acuminata* (Staudinger)	clover gelechiid
コミドリメクラガメ（コアオメクラガメ）	*Apolygus lucorum* (Meyer-Dür)	
コンドウヒゲナガアブラムシ	*Acyrthosiphon kondoi* Shinji	blue alfalfa aphid
サッポロウンカ	*Kakuna sapporonis* (Matsumura)	Japanese rice planthopper
サツマイモネコブセンチュウ	*Meloidogyne incognita* (Kofoid et White) Chitwood	southern root-knot nematode
サビヒョウタンゾウムシ	*Scepticus griseus* (Roelofs)	
シバクキハナバエ	*Atherigona reversura* Villeneuve	Bermudagrass stem maggot
シバネコブセンチュウ	*Meloidogyne graminis* (Sledge et Golden) whitehead	grass root-knot nematode
シバミノガ	*Nipponopsyche fuscescens* Yazaki	lawn grass bagworm
ジャワネコブセンチュウ	*Meloidogyne javanica* (Treub) Chitwood	Javanese root-knot nematode
シラホシカメムシ	*Eysarcoris ventralis* (Westwood)	whitespotted bug
シロオビノメイガ	*Hymenia recurvalis* (Fabricius)	beet webworm, Hawaiian beet webmoth
シロシタヨトウ	*Sarcopolia illoba* (Butler)	mulberry caterpillar
シロミャクイチモンジヨコバイ	*Paramesodes albinervosus* (Matsumura)	
シロモンヤガ	*Xestia c-nigrum* (Linnaeus)	cutworm
スゲドクガ	*Laelia coenosa sangaica* Moore	sedge tussock moth
スジキリヨトウ	*Spodoptera depravata* (Butler)	lawn grass cutworm

昆虫および動物名（和名順）

スジコガネ	*Mimela testaceipes* (Motschulsky)	lineate chafer
セジロウンカ	*Sogatella furcifera* (Horváth)	whitebacked rice planthopper
セスジナガウンカ	*Stenocranus minutus* (Fabricius)	
ソラマメヒゲナガアブラムシ	*Megoura crassicauda* Mordvilko	
ソルガムタマバエ（ソルゴータマバエ）	*Contarinia sorghicola* (Coquillett)	sorghum midge
タネバエ	*Delia platura* (Meigen)	seedcorn maggot
タマナヤガ	*Agrotis ipsilon* (Hufnagel)	black cutworm, dark sword grass moth
チャイロカメムシ	*Eurygaster koreana* Wagner	
ツマグロイナゴ（ツチバッタ）	*Mecostethus magister* (Rehn)	
ツマグロヨコバイ	*Nephotettix cincticeps* (Uhler)	green rice leafhopper
ツメクサガ	*Heliothis maritima* (Graslin)	flax budworm
ツメクサクダアザミウマ	*Haplothrips niger* (Osborn)	red clover thrips, daisy thrips
ツメクサシロカイガラムシ	*Aulacaspis trifolium* Takagi	clover root scale
ツメクササタコゾウムシ	*Hypera nigrirostris* (Fabricius)	
ツメクサベニマルアブラムシ	*Nearctaphis bakeri* (Cowen)	clover aphid, shortbreaked clover aphid
ツヤコガネ	*Anomala lucens* Ballion	shiny chafer
テンクロアツバ	*Rivula sericealis* (Scopoli)	leguminose rivula
ドウガネブイブイ	*Anomala cuprea* Hope	cupreous chafer
トウモロコシアブラムシ	*Rhopalosiphum maidis* (Fitch)	corn leaf aphid, corn aphid, cereal leaf aphid
トウモロコシウンカ	*Peregrinus maidis* (Ashmead)	corn planthopper
トガリヨコバイ	*Doratulina producta* (Matsumura)	
トゲシラホシカメムシ	*Eysarcoris parvus* Uhler	whitespotted spined bug
トノサマバッタ	*Locusta migratoria* Linnaeus	Asiatic locust, oriental migratory locust
ドバト	*Columba livia* Linnaeus	feral pigeon
トバヨコバイ	*Recilia tobai* (Matsumura)	

[236] 昆虫および動物名（和名順）

トビイロウンカ	*Nilaparvata lugens* (Stål)	brown rice planthopper
トビイロハゴロモ	*Mimophantia maritima* Matsumura	
トビイロムナボソコメツキ	*Agriotes ogurae fuscicollis* Miwa	barley wireworm
ドブネズミ	*Rattus norvegicus* Berkenhout	Norway rat, brown rat
ナガグロメクラガメ	*Adelphocoris suturalis* (Jakovlev)	
ナガチャコガネ	*Heptophylla picea* Motschulsky	
ナシケンモン	*Viminia rumicis* (Linnaeus)	sorrel cutworm
ナミハダニ	*Tetranychus urticae* Koch	two-spotted spider mite
ナミデセンセンチュウ	*Helicotylenchus dihystera* (Cobb) Sher	Cobb spiral nematode
ナモグリバエ	*Chromatomyia horticola* (Goureau)	garden pea leafminer
ニセナミハダニ	*Tetranychus cinnabarinus* (Boisduval)	carmine spider mite
ノウサギ	*Lepus brachyurus* Temminck	Japanese hare
ノコギリネグサレセンチュウ	*Pratylenchus crenatus* Loof	
ノハラナメクジ	*Deroceras reticulatum* (Müller)	grey field slug, field slug, netted slug
ハシブトガラス	*Corvus macrorhynchos* Wagler	jungle crow
ハシボソガラス	*Corvus corone* Linnaeus	carrion crow
ハスモンヨトウ	*Spodoptera litura* (Fabricius)	common cutworm, cluster caterpillar, cotton leafworm
ハタネズミ	*Microtus montebelli* Milne-Edwards	Japanese field vole
ハナムグリ	*Eucetonia pilifera* (Motschulsky)	
ハマベアワフキ	*Aphrophora maritima* Matsumura	
ヒエウンカ	*Sogatella panicicola* (Ishihara)	panicum planthopper
ヒエドロオイムシ（アワクビホソハムシ）	*Oulema dilutipes* (Fairmaire)	
ヒエノアブラムシ	*Melanaphis sacchari* (Zehntner)	cane aphid, sugarcane aphid

ヒゲナガメクラガメ	*Adelphocoris lineolatus* (Goeze)	alfalfa plant bug, lucerne plant bug
ヒゲブトアザミウマ	*Chirothrips manicatus* (Haliday)	timothy thrips
ヒシウンカ	*Pentastiridius apicalis* (Uhler)	rhombic planthopper
ヒシバッタ	*Tetrix japonica* (Bolivar)	
ヒメアオズキンヨコバイ	*Batracomorphus diminutus* (Matsumura)	
ヒメアシナガコガネ	*Ectinohoplia obducta* (Motshulsky)	
ヒメキバネサルハムシ (ダイズヒメサルハムシ)	*Pagria signata* (Motschulsky)	bean leaf beetle
ヒメクサキリ	*Homorocoryphus jezoensis* (Matsumura et Shiraki)	
ヒメコガネ	*Anomala rufocuprea* Motshulsky	soybean beetle
ヒメサクラコガネ	*Anomala geniculata* (Motschulsky)	
ヒメトビウンカ	*Laodelphax striatellus* (Fallén)	small brown planthopper
ヒメビロウドコガネ	*Maladera orientalis* (Motschulsky)	
ヒメフタテンヨコバイ	*Macrosteles striifrons* Anufriev	
フタスジメクラガメ	*Stenotus binotatus* (Fabricius)	timothy plant bug
フタスジヨコバイ	*Futasujinus candidus* (Matsumura)	
フタテンヨコバイ	*Macrosteles fascifrons* (Stål)	aster leafhopper
ブチヒゲカメムシ	*Dolycoris baccarum* (Linnaeus)	sloe bug
ベントグラスセンチュウ	*Anguina agrostis* (Steinbuck) Filipjev	bentgrass nematode, grass nematode
ホシアワフキ	*Aphrophora stictica* Matsumura	
ホシササキリ	*Conocephalus maculatus* (Le Guillou)	
ホソハリカメムシ	*Cletus punctiger* (Dallas)	
ホタルハムシ (クロバネアシナガハムシ)	*Monolepta dichroa* Harold	
マイマイガ (ブランコケムシ)	*Lymantria dispar* (Linnaeus)	gypsy moth

昆虫および動物名（和名順）

和名	学名	英名
マダラヨコバイ	*Psammotettix striatus* (Linnaeus)	
マメアブラムシ	*Aphis craccivora* Koch	cowpea aphid
マメコガネ	*Popillia japonica* Newmen	Japanese beetle
マメドクガ	*Cifuna locuples confusa* (Bremer)	bean tussock moth
マメハンミョウ	*Epicauta gorhami* Marseul	
マメヒメサヤムシガ（マメサヤヒメハマキ）	*Matsumuraeses phaseoli* (Matsumura)	soybean podworm
マルアワフキ	*Lepyronia grossa* Uhler	
マルクビクシコメツキ	*Melanotus fortnumi* Candeze	sweetpotato wireworm
マルシラホシカメムシ	*Eysarcoris guttiger* (Thunberg)	
ミドリヒメヨコバイ	*Edwardsiana flavescens* (Fabricius)	small green leafhopper
ミドリヨコバイ	*Elymana sulphurella* (Zetterstedt)	
ミナミネグサレセンチュウ	*Pratylenchus coffeae* (Zimmermann) Filipjev et Schuurmans Stekhoven	coffee root-lesion nematode
ムギキイロハモグリバエ	*Cerodontha denticornis* (Panzer)	barley leafminer
ムギクビレアブラムシ	*Rhopalosiphum padi* (Linnaeus)	oat bird-cherry aphid, bird-cherry aphid, oat aphid, wheat aphid
ムギシリトゲハモグリバエ	*Agromyza cinerascens* Macquart	
ムギダニ	*Penthaleus major* (Dugès)	winter grain mite, blue oat mite
ムギネグサレセンチュウ	*Pratylenchus neglectus* (Rensch) Filipjev et Schuurmans Stekhoven	California root-lesion nematode
ムギノミハムシ（ムギヒサゴトビハムシ）	*Chaetocnema cylindrica* (Baly)	barley flea beetle
ムギヒゲナガアブラムシ	*Sitobion akebiae* (Shinji)	
ムギメクラガメ	*Stenodema calcaratum* (Fallén)	wheat leaf bug
ムツテンヨコバイ	*Macrosteles sexnotatus* (Fallén)	
モモアカアブラムシ	*Myzus persicae* (Sulzer)	green peach aphid, peach-potato aphid
モリウスピンセンチュウ	*Paratylenchus morius* Yokoo	

モロコシネグサレセンチュウ	*Pratylenchus zeae* Graham	corn root-lesion nematode
モンキアワフキ	*Aphrophora flavomaculata* Matsumura	
モンキチョウ	*Colias erate* poliographus Motschulsky	oriental clouded yellow, eastern pale clouded yellow
ヤノトガリヨコバイ	*Yanocephalus yanonis* (Matsumura)	
ヤノハモグリバエ	*Agromyza yanonis* (Matsumura)	barley leafminer
ヨツテンヨコバイ	*Macrosteles quadrimaculatus* (Matsumura)	
ヨツモンヒメヨコバイ	*Empoascanara limbata* (Matsumura)	
ヨツモンヨコバイ	*Cicadula quadrinotata* (Fabricius)	
ヨトウガ	*Mamestra brassicae* (Linnaeus)	cabbage armyworm
ヨモギエダシャク	*Ascotis selenaria* (Denis et Schiffermüller)	mugwort looper

昆虫および動物名 （学名順）

学名	和名	英名
Acantholeucania loreyi (Duponchel)	クサシロキヨトウ	
Acusta despecta sieboldiana (Pfeiffer)	ウスカワマイマイ	
Acyrthosiphon kondoi Shinji	コンドウヒゲナガアブラムシ	blue alfalfa aphid
Acyrthosiphon pisum (Harris)	エンドウヒゲナガアブラムシ	pea aphid
Adelphocoris lineolatus (Goeze)	ヒゲナガメクラガメ	alfalfa plant bug, lucerne plant bug
Adelphocoris suturalis (Jakovlev)	ナガグロメクラガメ	
Aeschyteles maculatus (Fieber)	アカヒメヘリカメムシ	carrot bug
Agriotes ogurae fuscicollis Miwa	トビイロムナボソコメツキ	barley wireworm
Agromyza cinerascens Macquart	ムギシリトゲハモグリバエ	
Agromyza yanonis (Matsumura)	ヤノハモグリバエ	barley leafminer
Agrotis ipsilon (Hufnagel)	タマナヤガ	black cutworm, dark sword grass moth
Agrotis segetum (Denis et Schiffermüller)	カブラヤガ	cutworm, turnip moth
Anaphothrips obscurus (Müller)	クサキイロアザミウマ	grass thrips
Anguina agrostis (Steinbuck) Filipjev	ベントグラスセンチュウ	bentgrass nematode, grass nematode
Anomala cuprea Hope	ドウガネブイブイ	cupreous chafer
Anomala geniculata (Motschulsky)	ヒメサクラコガネ	
Anomala lucens Ballion	ツヤコガネ	shiny chafer
Anomala rufocuprea Motshulsky	ヒメコガネ	soybean beetle
Aphis craccivora Koch	マメアブラムシ	cowpea aphid
Aphrophora flavomaculata Matsumura	モンキアワフキ	
Aphrophora maritima Matsumura	ハマベアワフキ	
Aphrophora stictica Matsumura	ホシアワフキ	
Apolygus lucorum (Meyer-Dür)	コミドリメクラガメ（コアオメクラガメ）	

昆虫および動物名（学名順）

Apolygus spinolai (Meyer-Dur)   ウスミドリメクラガメ（ツマグロアオメクラガメ）
Armadillidium vulgare Latreille   オカダンゴムシ   pillbug
Ascotis selenaria (Denis et Schiffermüller)   ヨモギエダシャク   mugwort looper
Atherigona reversura Villeneuve   シバクキハナバエ   Bermudagrass stem maggot
Atrachya menestriesi (Faldermann)   ウリハムシモドキ   false melon beetle
Aulacaspis trifolium Takagi   ツメクサシロカイガラムシ   clover root scale
Aulacophora femoralis (Motschulsky)   ウリハムシ   cucurbit leaf beetle
Balclutha punctata (Fabricius)   カスリヨコバイ
Balclutha rubinervis (Matsumura)   アカカスリヨコバイ
Balclutha viridis (Matsumura)   ウスバミドリヨコバイ
Basilepta fulvipes (Motschulsky)   アオバネサルハムシ（キアシサルハムシ）
Batracomorphus diminutus (Matsumura)   ヒメアオズキンヨコバイ
Batracomorphus mundus (Matsumura)   アオズキンヨコバイ
Bourletiella hortensis (Fitch)   キボシマルトビムシ   garden springtail
Bruchophagus gibbus (Boheman)   クローバタネコバチ   clover seed chalcid
Bryobia praetiosa koch   クローバハダニ   clover mite
Cerodontha denticornis (Panzer)   ムギキイロハモグリバエ   barley leafminer
Cetema cereris (Fallén)   エゾスキガタキモグリバエ
Chaetocnema cylindrica (Baly)   ムギノミハムシ（ムギヒサゴトビハムシ）   barley flea beetle
Chirothrips manicatus (Haliday)   ヒゲブトアザミウマ   timothy thrips
Chromatomyia horticola (Goureau)   ナモグリバエ   garden pea leafminer
Cicadella viridis (Linnaeus)   オオヨコバイ
Cicadula quadrinotata (Fabricius)   ヨツモンヨコバイ
Cifuna locuples confusa (Bremer)   マメドクガ   bean tussock moth
Cletus punctiger (Dallas)   ホソハリカメムシ
Colias erate poliographus Motschulsky   モンキチョウ   oriental clouded yellow, eastern pale

		clouded yellow
Columba livia Linnaeus	ドバト	feral pigeon
Conocephalus maculatus (Le Guillou)	ホシササキリ	
Conocephalus chinensis (Redtenbacher)	ウスイロササキリ	
Conocephalus gladiatus (Redtenbacher)	オナガササキリ	
Conocephalus japonicus (Redtenbacher)	コバネササキリ	
Contarinia sorghicola (Coquillett)	ソルガムタマバエ（ソルゴータマバエ）	sorghum midge
Corvus corone Linnaeus	ハシボソガラス	carrion crow
Corvus macrorhynchos Wagler	ハシブトガラス	jungle crow
Creontiades pallidifer (Walker)	アカホシメクラガメ	
Delia platura (Meigen)	タネバエ	seedcorn maggot
Deroceras reticulatum (Müller)	ノハラナメクジ	grey field slug, field slug, netted slug
Dichomeris acuminata (Staudinger)	コフサキバガ	clover gelechiid
Dolycoris baccarum (Linnaeus)	ブチヒゲカメムシ	sloe bug
Doratulina grandis (Matsumura)	オオトガリヨコバイ	
Doratulina producta (Matsumura)	トガリヨコバイ	
Eana (Ablabia) argentana (Clerck)	ギンムジハマキ	
Ectinohoplia obducta (Motshulsky)	ヒメアシナガコガネ	
Ectinus sericeus (Candeze)	カバイロコメツキ	wheat wireworm
Edwardsiana flavescens (Fabricius)	ミドリヒメヨコバイ	small green leafhopper
Elymana sulphurella (Zetterstedt)	ミドリヨコバイ	
Empoascanara limbata (Matsumura)	ヨツモンヒメヨコバイ	
Endoclyta excrescens (Butler)	コウモリガ	swift moth
Epicauta gorhami Marseul	マメハンミョウ	
Eucetonia pilifera (Motschulsky)	ハナムグリ	
Euconocephalus thunbergi (Stål)	クビキリギリス	
Eurygaster koreana Wagner	チャイロカメムシ	
Exitianus indicus (Distant)	クロミャクイチモンジヨコバイ	
Eysarcoris guttiger (Thunberg)	マルシラホシカメムシ	

昆虫および動物名（学名順）

Eysarcoris lewisi (Distant)	オオトゲシラホシカメムシ	
Eysarcoris parvus Uhler	トゲシラホシカメムシ	whitespotted spined bug
Eysarcoris ventralis (Westwood)	シラホシカメムシ	whitespotted bug
Frankliniella tenuicornius (Uzel)	カホンカハナアザミウマ	
Futasujinus candidus (Matsumura)	フタスジヨコバイ	
Haplothrips aculeatus (Fabricius)	イネクダアザミウマ	rice aculeated thrips
Haplothrips niger (Osborn)	ツメクサクダアザミウマ	red clover thrips, daisy thrips
Helicotylenchus dihystera (Cobb) Sher	ナミラセンセンチュウ	Cobb spiral nematode
Heliothis maritima (Graslin)	ツメクサガ	flax budworm
Heptophylla picea Motschulsky	ナガチャコガネ	
Heterodera trifolii Goffart	クローバシストセンチュウ	clover cyst nematode
Holotrichia kiotonensis Brenske	クロコガネ	balck chafer
Homorocoryphus jezoensis (Matsumura et Shiraki)	ヒメクサキリ	
Hoplia communis Waterhouse	アシナガコガネ	
Hydrellia griseola (Fallén)	イネミギワバエ（イネヒメハモグリバエ）	rice leafminer
Hymenia recurvalis (Fabricius)	シロオビノメイガ	beet webworm, Hawaiian beet webmoth
Hypera nigrirostris (Fabricius)	ツメクサタコゾウムシ	
Hypera postica (Gyllenhal)	アルファルファタコゾウムシ	alfalfa weevil
Hypera punctata (Fabricius)	オオタコゾウムシ	clover leaf weevil
Kakuna sapporonis (Matsumura)	サッポロウンカ	Japanese rice planthopper
Laelia coenosa sangaica Moore	スゲドクガ	sedge tussock moth
Laodelphax striatellus (Fallén)	ヒメトビウンカ	small brown planthopper
Leptocorisa chinensis (Dallas)	クモヘリカメムシ	rice bug
Lepus brachyurus Temminck	ノウサギ	Japanese hare
Lepus timidus ainu Barrett-Hamilton	エゾユキウサギ	Hokkaido hare
Lepyronia grossa Uhler	マルアワフキ	
Locusta migratoria Linnaeus	トノサマバッタ	Asiatic locust, oriental migratory locust

Lymantria dispar (Linnaeus)	マイマイガ (ブランコケムシ)	gypsy moth
Macrosteles fascifrons (Stål)	フタテンヨコバイ	aster leafhopper
Macrosteles quadrimaculatus (Matsumura)	ヨツテンヨコバイ	
Macrosteles sexnotatus (Fallén)	ムツテンヨコバイ	
Macrosteles striifrons Anufriev	ヒメフタテンヨコバイ	
Maladera castanea (Arrow)	アカビロウドコガネ	Asiatic garden beetle
Maladera orientalis (Motschulsky)	ヒメビロウドコガネ	
Mamestra brassicae (Linnaeus)	ヨトウガ	cabbage armyworm
Matsumuraeses phaseoli (Matsumura)	マメヒメサヤムシガ (マメサヤヒメハマキ)	soybean podworm
Mecostethus magister (Rehn)	ツマグロイナゴ (ツチバッタ)	
Megoura crassicauda Mordvilko	ソラマメヒゲナガアブラムシ	
Melanaphis sacchari (Zehntner)	ヒエノアブラムシ	cane aphid, sugarcane aphid
Melanotus erythropygus Candeze	コガタクシコメツキ	
Melanotus fortnumi Candeze	マルクビクシコメツキ	sweetpotato wireworm
Melanotus legatus Candeze	クシコメツキ	
Melanotus senilis Candeze	クロクシコメツキ	
Meloidogyne arenaria (Neal) Chitwood	アレナリアネコブセンチュウ	peanut root-knot nematode
Meloidogyne graminis (Sledge et Golden) whitehead	シバネコブセンチュウ	grass root-knot nematode
Meloidogyne hapla Chitwood	キタネコブセンチュウ	northern root-knot nematode
Meloidogyne incognita (Kofoid et White) Chitwood	サツマイモネコブセンチュウ	southern root-knot nematode
Meloidogyne javanica (Treub) Chitwood	ジャワネコブセンチュウ	Javanese root-knot nematode
Microtus montebelli Milne-Edwards	ハタネズミ	Japanese field vole
Mimela testaceipes (Motschulsky)	スジコガネ	lineate chafer
Mimophantia maritima Matsumura	トビイロハゴロモ	
Mogera wogura (Temminck)	アズマモグラ	Japanese mole

昆虫および動物名（学名順）

Monolepta dichroa Harold	ホタルハムシ（クロバネアシナガハムシ）	
Mythimna separata (Walker)	アワヨトウ	oriental armyworm, rice armyworm, rice ear-cutting caterpillar
Myzus persicae (Sulzer)	モモアカアブラムシ	green peach aphid, peach-potato aphid
Nearctaphis bakeri (Cowen)	ツメクサベニマルアブラムシ	clover aphid, shortbreaked clover aphid
Nephotettix cincticeps (Uhler)	ツマグロヨコバイ	green rice leafhopper
Nephrotoma aculeata (Loew)	カラフトアオホソガガンボ	
Nephrotoma cornicina (Linnaeus)	エゾホソガガンボ	
Nephrotoma virgata (Coquillett)	キイロホソガガンボ（キスジガガンボ）	
Nilaparvata lugens (Stål)	トビイロウンカ	brown rice planthopper
Nipponopsyche fuscescens Yazaki	シバミノガ	lawn grass bagworm
Oedaleus infernalis de Saussure	クルマバッタモドキ	
Ostrinia furnacalis (Guenée)	アワノメイガ	oriental corn borer
Oulema dilutipes (Fairmaire)	ヒエドロオイムシ（アワクビホソハムシ）	
Oxya yezoensis Shiraki	コバネイナゴ	rice grasshopper
Oxycetonia jucunda (Faldermann)	コアオハナムグリ	citrus flower chafer
Pagaronia guttigera (Uhler)	クワキヨコバイ	yellow mulberry leafhopper
Pagria signata (Motschulsky)	ヒメキバネサルハムシ（ダイズヒメサルハムシ）	bean leaf beetle
Paragabara flavomacula (Oberthür)	キボシアツバ	
Paramesodes albinervosus (Matsumura)	シロミャクイチモンジヨコバイ	
Parapleurus alliaceus (Germar)	イナゴモドキ	
Paratylenchus morius Yokoo	モリウスピンセンチュウ	
Pentastiridius apicalis (Uhler)	ヒシウンカ	rhombic planthopper

昆虫および動物名（学名順） [247]

Penthaleus major (Dugès)	ムギダニ	winter grain mite, blue oat mite"
Peregrinus maidis (Ashmead)	トウモロコシウンカ	corn planthopper
Perkinsiella saccharicida Kirkaldy	クロフツノウンカ	sugarcane planthopper, sugarcane leafhopper
Phasianus colchicus Linnaeus	キジ	common pheasant
Phragmatobia fuliginosa japonica Rothschild	アマヒトリ	flax arctid, ruby tiger
Piezodorus hybneri (Gmelin)	イチモンジカメムシ	
Pleuroptya ruralis (Scopoli)	ウコンノメイガ	bean webworm
Plutella xylostella (Linnaeus)	コナガ	diamondback moth, cabbage moth
Popillia japonica Newmen	マメコガネ	Japanese beetle
Pratylenchus coffeae (Zimmermann) Filipjev et Schuurmans Stekhoven	ミナミネグサレセンチュウ	coffee root-lesion nematode
Pratylenchus crenatus Loof	ノコギリネグサレセンチュウ	
Pratylenchus neglectus (Rensch) Filipjev et Schuurmans Stekhoven	ムギネグサレセンチュウ	California root-lesion nematode
Pratylenchus penetrans (Cobb) Filipjev et Schuurmans Stekhoven	キタネグサレセンチュウ	Cobb root-lesion nematode
Pratylenchus vulnus Allen et Jensen	クルミネグサレセンチュウ	walnut root-lesion nematode
Pratylenchus zeae Graham	モロコシネグサレセンチュウ	corn root-lesion nematode
Psammotettix striatus (Linnaeus)	マダラヨコバイ	
Rattus norvegicus Berkenhout	ドブネズミ	Norway rat, brown rat
Recilia oryzae (Matsumura)	イネマダラヨコバイ	
Recilia tobai (Matsumura)	トバヨコバイ	
Rhopalosiphum maidis (Fitch)	トウモロコシアブラムシ	corn leaf aphid, corn aphid, cereal leaf aphid
Rhopalosiphum padi (Linnaeus)	ムギクビレアブラムシ	oat bird-cherry aphid, bird-cherry aphid, oat aphid, wheat aphid
Rivula sericealis (Scopoli)	テンクロアツバ	leguminose rivula

昆虫および動物名 （学名順）

Sarcopolia illoba (Butler)	シロシタヨトウ	mulberry caterpillar
Scepticus griseus (Roelofs)	サビヒョウタンゾウムシ	
Scleroracus flavopictus (Ishihara)	キマダラヒロヨコバイ	
Sesamia inferens (Walker)	イネヨトウ	pink borer
Sitobion akebiae (Shinji)	ムギヒゲナガアブラムシ	
Sitona hispidulus (Fabricius)	ケチビコフキゾウムシ	clover root curculio
Sminthurus viridis (Linnaeus)	キマルトビムシ	lucerne flea
Sogatella furcifera (Horvath)	セジロウンカ	whitebacked rice planthopper
Sogatella panicicola (Ishihara)	ヒエウンカ	panicum planthopper
Spilosoma lubricipedum (Linnaeus)	キハラゴマダラヒトリ	
Spodoptera depravata (Butler)	スジキリヨトウ	lawn grass cutworm
Spodoptera litura (Fabricius)	ハスモンヨトウ	common cutworm, cluster caterpillar, cotton leafworm
Stenocranus minutus (Fabricius)	セスジナガウンカ	
Stenodema calcaratum (Fallén)	ムギメクラガメ	wheat leaf bug
Stenotus binotatus (Fabricius)	フタスジメクラガメ	timothy plant bug
Stenotus rubrovittatus (Matsumura)	アカスジメクラガメ	sorghum plant bug
Streptopellia orientalis (Latham)	キジバト	rufous turtle dove
Succinea lauta Gould	オカモノアラガイ	
Teleogryllus emma (Ohmachi et Matsuura)	エンマコオロギ	field cricket
Tetranychus cinnabarinus (Boisduval)	ニセナミハダニ	carmine spider mite
Tetranychus kanzawai Kishida	カンザワハダニ	Kanzawa spider mite
Tetranychus urticae Koch	ナミハダニ	two-spotted spider mite
Tetrix japonica (Bolivar)	ヒシバッタ	
Tipula subcunctans Alexander	ウスイロガガンボ	
Trigonotylus coelestialium (Kirkaldy)	アカヒゲホソミドリメクラガメ	rice leaf bug
Viminia rumicis (Linnaeus)	ナシケンモン	sorrel cutworm
Xestia c-nigrum (Linnaeus)	シロモンヤガ	cutworm
Xylena fumosa (Butler)	アヤモクメキリガ	rape catapillar
Yanocephalus yanonis (Matsumura)	ヤノトガリヨコバイ	

| R |〈学術著作権協会へ複写権委託〉 |

2000 　　2000年6月15日　改訂第1版発行

改訂 草地学用語集		
著者との申 し合せによ り検印省略	編　集　者	日 本 草 地 学 会
	発　行　者	株式会社　養 賢 堂 代表者　及川　清
©著作権所有		
本体 3600 円	印　刷　者	星野精版印刷株式会社 責任者　星野恭一郎

発行所　〒113-0033 東京都文京区本郷5丁目30番15号
　　　　株式
会社 養賢堂　電　話 東京(03)3814-0911　振替00120
FAX 東京(03)3812-2615　7-25700

ISBN4-8425-0060-3 C3061

PRINTED IN JAPAN　　　　　製本所　板倉製本印刷株式会社

本書の無断複写は、著作権法上での例外を除き、禁じられています。
本書からの複写承諾は、学術著作権協会(〒107-0052東京都港区赤坂
9-6-41乃木坂ビル、電話03-3475-5618、FAX03-3475-5619)から得て
下さい。